Augustine Thompson O.P. is Assistant
Professor of Medieval Christianity at the
University of Oregon, Eugene, and himself a
preacher.

Revival Preachers and Politics in Thirteenth-Century Italy

The Great Devotion of 1233

AUGUSTINE THOMPSON, O.P.

CLARENDON PRESS · OXFORD

1992

Oxford University Press, Walton Street, Oxford OX2 6DP
Oxford   New York   Toronto
Delhi   Bombay   Calcutta   Madras   Karachi
Petaling Jaya   Singapore   Hong Kong   Tokyo
Nairobi   Dar es Salaam   Cape Town
Melbourne   Auckland
and associated companies in
Berlin   Ibadan

Oxford is a trademark of Oxford University Press

Published in the United States
by Oxford University Press, New York

British Library Cataloguing in Publication Data
Data available

Library of Congress Cataloging in Publication Data
Thompson, Augustine.
Revival preachers and politics in thirteenth-century Italy: the
great devotion of 1233 / Augustine Thompson.
p.   cm.
Includes bibliographical references and index.
1. Preaching—Italy—History.   2. Revivals—Italy—History.
3. Christianity and politics—History.   I. Title.
BV4208.I8T48   1992
251'.00945'09022—dc20   92-15169
ISBN 0-19-820287-3

Typeset by Best-set Typesetter Ltd., Hong Kong
Printed and bound in
Great Britain by Bookcraft Ltd,
Midsomer Norton, Bath

ACKNOWLEDGEMENTS

A HOMILETICS instructor once told me that a preacher should begin with a story and then develop from that story the points he wishes to make in the sermon. Now, as I reread this work, I find that I have unconsciously done just that. This work first recounts the remarkable series of events in north Italy called the 'Great Devotion of 1233', and then, by focusing on five aspects of that revival, reconstructs the essential components of the 'Ancient Way of Preaching' which that revival typified. This book could have been written in other ways, and it had a vastly different form when I first presented it as a doctoral dissertation at the University of California, Berkeley. The research for that thesis provided the core of this work. Then, drawing on the insightful comments of friends and readers, I completely reorganized the structure of the narrative and added three new chapters. There is, in this book, hardly a paragraph from the thesis that has not undergone some revision or correction thanks to the suggestions of those who read the typescript.

In undertaking the preparation of this text, I have enjoyed much encouragement from my teacher, Professor Robert Brentano of Berkeley. Had it not been for him, I would probably never have revised this work for publication. Professor Brentano once divided scholarly works into two types: the small closed book, and the large open book.[1] The latter are much to be preferred to the former because they carry the reader beyond them into a greater world. The nucleus of this work is a small book, on a small region (northern Italy), focusing on a small period of time (January to November, 1233). I hope that, even if it is a small book, it might still be an open one and one that will transport the reader into the much larger world of thirteenth-century preaching and politics. If it can do this, Professor Brentano deserves most of the credit.

During my writing and research I have incurred a debt of gratitude to many others as well. For hospitality and the peace to work, I am indebted to the Dominican communities in Bologna, Siena, Oakland (California), Eugene (Oregon), and Ashland (Oregon). The staffs of the Biblioteca Comunale of Siena, the Centro di Documentazione, the Biblioteca Universitaria and the Biblioteca dell'Archiginnasio of Bologna, the

---

[1] Robert Brentano, *Two Churches: England and Italy in the Thirteenth Century*, 2nd edn. (Berkeley, Calif., 1988), 378.

Centro per Diritto Comparato of Florence, the Biblioteca Antoniana of Padua, and the Biblioteca Conventuale Domenicana of Bologna all treated me with kindness and compassion during my research, something for which I am profoundly grateful.

This work owes its very existence to Professor James Gordley of the University of California Law School. He has aided me not only by his learning but also by his unstinting generosity and hospitality. His wife, Barbara Pike Gordley, a finer historian than I, has saved me from an uncounted number of humiliating blunders. Finally, I am very grateful to the members of my own Dominican Province and its Father Provincial, John Flannery, for their constant support and generosity.

To thank by name each of the many others to whom I am indebted is impossible, so I shall say, as Salimbene de Adam said of his preachers: Et multi alii, quos vidi et cognovi, quorum memoria sit Deo! Amen.

The Translation of St Dominic   A. T.
Eugene, Oregon
1991

CONTENTS

List of Tables and Maps   viii

A Note on Proper Names and Latin Texts   ix

Abbreviations   x

Introduction   I

PART I: THE GREAT DEVOTION OF 1233   27

1. The Coming of the Preachers   29
2. The Great Prophet of Bologna   45
3. Peace Campaigns in the North   63

PART II: REVIVALISM AND POLITICS   81

4. The Making of a Revival   83
5. The Revivalist as Miracle-Worker   110
6. The Revivalist as Peace-Maker   136
7. The Revivalist as Arbiter   157
8. The Revivalist as Legislator   179

Conclusion: Revivals and Politics   205

Appendix: Borselli's Unedited Account of 1233   219

Bibliography of Works Cited   221

Index   239

LIST OF TABLES AND MAPS

TABLES

1. Comparison of Miracles by Type 115
2. Gerard's Legislation at Parma 183

MAPS

1. Northern Italy 30
2. The *Contado* of Bologna 46

A NOTE ON PROPER NAMES AND LATIN TEXTS

The translation of medieval names, especially more obscure ones, presents countless problems. In this work I have, when at all possible, given the names of medieval figures in English form. When no English form exists, I have made do with the Italian one. Sometimes, when English-speakers know a figure under a foreign form, I have kept it. This was the case for several chroniclers and legists. For the greatest preacher of 1233, I have chosen the form John 'of Vicenza' over the commonly accepted John 'of Schio.'[1] Popular as the latter form is, it was unknown before the fourteenth century. To himself and to his contemporaries he was always Giovanni da Vicenza or Giovanni di Bologna, the latter used because, as Ambrose Sansedoni tells us, Bologna was where he worked most of his miracles.[2] Ambrose, his near contemporary and fellow Dominican, would have known the common usage. The general rule has been to choose the form that seemed least likely to distract the reader, rather than to impose arbitrary uniformity.

When quoting Latin texts, I have preserved the punctuation, capitalization, and spelling of the editions, even when this does not conform to contemporary usage. In quotations from manuscripts, I have modernized the punctuation and capitalization, but preserved the spelling.

[1] The first to link John with Schio was the Vicenzan chronicler Godi, 10, who called him 'Ioannes de Scledo'. The theory that he belonged to the family of the counts of Schio appeared in the 17th cent., when Lodovico da Schio adopted John as a distant relative in his *Vita del beato Giovanni da Schio dell'Ordine de' Predicatori*, Vicenza, Biblioteca Comunale MS E. 2. 26.

[2] Ambrose Sansedoni, *Sermones de Tempore*, Siena, Biblioteca Comunale MS T. IV. 7, fol. 133^r. On this manuscript, see Thomas Kaeppeli, 'Le prediche del b. Ambrogio Sansedoni da Siena', *AFP* 38 (1968), 5–12.

ABBREVIATIONS

Citations from the Roman and Canon Law are given in the usual form: C., I., D., and Nov. for the Code, Institutes, Digest, and Novels of the *Corpus Iuris Civilis*; D. 1 c. 1 for the first part of Gratian's *Decretum*; C. 1 q. 1 c. 1 for the second part; D. 1 c. 1, *De poen.* for Gratian's tract *De Poenitentia*; X for the *Decretals of Gregory IX*.

Act. Can. Dom.	*Acta Canonizationis Sancti Dominici*, ed. Angelus Walz (MOFPH 16: MHD 2; Rome, 1935).
ACG	*Acta Capitulorum Generalium Ordinis Praedicatorum* i, ed. Benedictus Maria Reichert (MOFPH 3; Rome, 1898).
ACPL	*Acta Capitulorum Provincialium Provinciae Lombardiae*, ed. Thomas Kaeppeli, in 'Acta Capitulorum Provinciae Lombardiae (1254–1293) et Lombardiae Inferioris (1309–1312)', *AFP* 11 (1941), 140–67.
ACPR	*Acta Capitulorum Provincialium Provincie Romanae (1243–1344)*, ed. Thomas Kaeppeli and Antoine Dondaine (MOFPH 20; Rome, 1941).
AF	Analecta Franciscana.
AFP	*Archivum Fratrum Praedicatorum*.
AIMA	*Antiquitates Italicae Medii Aevi*, ed. Lodovico Antonio Muratori (6 vols.; Milan, 1741).
Alessandrino	Gilsberto Alessandrino, Recuperato of Pietramala, Aldobrandino Papparoni, and Oldrado Visdomni, *Legenda Antica Beati Ambrosii Senensis*, ed. Daniel Papabroch (AS 9: Mar. III; Paris, 1865).
Ann. Bergom.	*Annales Bergomates 1156–1266*, ed. Oswald Holder-Egger (MGH.Ss. 31).
Ann. Brix.	*Annales Brixienses*, ed. Ludwig Bethmann, (MGH.Ss. 18).
Ann. Parm. Maj.	*Annales Parmenses Maiores*, ed. Georg Heinrich Pertz (MGH.Ss. 18).
Ann. Plac. Guelf.	*Annales Placentini Guelfi*, ed. Georg Heinrich Pertz (MGH.Ss. 18).
Ann. Vet. Mut.	*Annales Veteres Mutinenses ab Anno MCXXXI usque ad MCCCXXXVI* (RIS 11).

Ann. Vet. *Veron.*	*Annales Veteres Veronenses*, ed. Carlo Cipolla, *Archivio Veneto*, 9 (1875), 77–98.
AS	Acta Sanctorum.
Assidua	*Vita prima di s. Antonio o 'Assidua' (c.1232)*, ed. and trans. Vergilio Gamboso (Padua, 1981).
Auvray	*Les Registres de Grégoire IX: Recueil des bulles de ce pape*, ed. Lucien Auvray (3 vols.; Paris, 1896).
Bergamo Stat.	*Antiquae Collationes Statuti Veteris Civitatis Pergami*, ed. Giovanni Finarzi (Turin, 1876).
BISI	*Bollettino dell'Istituto Storico Italiano per il medio evo e Archivo muratoriano.*
BOFM	*Bullarium Franciscanum Romanorum Pontificum*, ed. Giovanni Giacinto Sbaraglia (7 vols.; Rome, 1759).
Bol. Armi Stat.	*Statuti delle Società del Popolo di Bologna*, i: *Società delle Armi*, ed. Augusto Gaudenzi (Rome, 1889).
Bol. Stat.	*Statuti di Bologna dall'anno 1245 all'anno 1267*, ed. Lodovico Frati (3 vols.; Bologna, 1869–77).
Bolognetti	*Cronaca detta dei Bolognetti*, ed. Albano Sorbelli (RIS[2] 18:1:2; Città di Castello, 1911).
BOP	*Bullarium Ordinis Fratrum Praedicatorum*, ed. Thomas Ripoll (8 vols.; Rome, 1729).
Borselli, *Cron.*	Girolamo Albertucci de' Borselli, *Cronica Gestorum ac Factorum Memorabilium Civitatis Bononie*, ed. Albano Sorbelli (RIS[2] 23:2; Città di Castello, 1912).
Borselli, *CMG*	Girolamo Albertucci de' Borselli, *Chronica Magistrorum Generalium Ordinis Fratrum Predicatorum*, Bologna, Biblioteca Universitaria MS 1999 (entry for 1233 edited in Appendix).
CCB	Corpus Chronicorum Bononiensium 2, ed. Albano Sorbelli (RIS[2] 18:1:2; Città di Castello, 1911).
Dom. Con.	Die Constitutionen des Predigerordens vom Jarhe 1228,' ed. H. Denifle, *Archiv für Literatur- und Kirchengeschichte des Mittelalters*, 1 (1885), 165–227.
EFRMAH	*École Française de Rome: Mélanges d'archéologie et d'histoire.*
Fiamma, *MF*	Galvano Fiamma, *Manipulus Florum* (RIS 11).
Fumagalli	Vito Fumagalli, 'In margine all' "Alleluia" del 1233', BISI 80 (1968), 257–72.
Godi	Antonio Godi, *Cronaca*, ed. Giovanni Soranzo (RIS[2] 8:1; Città di Castello, 1905).

H.-B.	Jean Louis Alphonse Huillard-Bréholles (ed.), *Historia Diplomatica Friderici Secundi* (7 vols. in 12; Paris, 1852–61).
Maurisio	Gerardo Maurisio, *Cronica Dominorum Ecelini et Alberici Fratrum de Romano*, ed. Giovanni Soranzo (RIS² 8:4; Città di Castello, 1914).
MEFR	*Mélanges de l'École Française de Rome: Moyen âge–temps modernes.*
MGH.Ss.	Monumenta Germaniae Historica: Scriptores
MHD	Monumenta Historica S. P. N. Dominici, (2 vols.; Rome, 1933–35).
MOFPH	Monumenta Ordinis Fratrum Praedicatorum Historica.
Monza Stat.	*Memorie storiche di Monza e sua corte*, ii: *Codice diplomatico*, ed. Anton-Francesco Frisi (Milan, 1794).
Mussi	Giovanni de' Mussi, *Chronicon Placentinum*, ed. Lodovico Antonio Muratori, (RIS 16).
Odofredus	Odofredus, *Lectura in Digesto Veteri* (2 vols.; 1550; repr. Bologna, 1970(?)).
Padua Stat.	*Statuti del comune di Padova dal secolo XII all'anno 1285*, ed. Andrea Gloria (Padua, 1872).
Parisio	Parisio of Cerea, *Annales Veronenses*, ed. Philipp Jaffé (MGH.Ss. 19).
Parma Stat.	*Statuta Communis Parmae Digesta Anno 1255*, (Parma, 1856).
PL	Patrologiae Cursus Completus, Series Latina, ed. Jean-Paul Migne (Paris, 1844–65).
Potthast	*Regestra Pontificum Romanorum inde ab Anno post Christum Natum 1198 ad Annum 1304*, ed. A. Potthast (2 vols.; Berlin, 1874–5).
Rainerius	Rainerius Perusinus, *Ars Notariae*, ed. Ludwig Wahrmund (Innsbruck, 1917).
Rampona	*Cronaca A detta volgarmente Rampona*, ed. Albano Sorbelli (RIS² 18:1:2; Città di Castello, 1911).
RI	*Die Regesten des Kaiserreichs unter Philipp, Otto IV., Friedrich II., Heinrich (VII.), Conrad IV., Heinrich Raspe, Wilhelm und Richard, 1198–1272*, ed. Julius Ficker and Eduard Winkelmann, Regista Imperii, v (3 vols. in 5; Innsbruck, 1881–1901).
Rich. Ang.	Richardus Anglicus, *Summa de Ordine Iudiciario*, ed. Ludwig Wahrmund (Innsbruck, 1915).

RIS	Rerum Italicarum Scriptores, ed. Lodovico Antonio Muratori (25 vols.; Milan, 1727).
RIS²	Rerum Italicarum Scriptores, 2nd edn. (34 vols.; Città di Castello, 1902–).
Rodenberg	*Epistolae Saeculi XIII ex Registris Pontificum Romanorum*, ed. Carl Rodenberg (3 vols.; Berlin, 1883–94).
Rolandino	Rolandino of Padua, *Cronica in Factis et circa Facta Marchie Trivixiane*, ed. Antonio Bonardi (RIS² 8:1; Città di Castello, 1905).
Rolandinus	Rolandinus Passagerii, *Summa Totius Artis Notariae* (2 vols.; Venice, 1588).
Salimbene	Salimbene de Adam, *Cronica*, ed. Oswald Holder-Egger (MGH.Ss. 32).
Savioli	Lodovico Vittorio Savioli, *Annali bolognesi* (6 vols.; Bassano, 1784–95).
SF	*Studi francescani.*
Sutter	Carl Sutter, *Johann von Vicenza und die italienische Friedensbewegung im Jahre 1233* (Freiburg, 1891).
Taegio	Ambrogio Taegio, *Legenda Beatissimi Petri Martyris ex Multis Legendis in Unum Compilata* (AS 12: Apr. III; Paris, 1867).
Tancred	Tancred of Bologna, *Ordo Iudiciarius*, ed. Friedrich Christian Bergmann, (Göttingen, 1965).
Th. Cant.	Thomas of Cantimpré, *Bonum Universale de Apibus*, ed. (in part) Joannes Baptista Sollerius (AS 28: Jul. I; Paris, 1867), 424–5.
Tonini	Luigi Tonini, *Della storia civile e sacra riminese*, iii (Rimini, 1862).
Trev. Stat.	*Gli statuti del comune di Treviso*, ii: *Statuti degli anni 1231—1233—1260–63*, ed. Giuseppe Liberali (Venice, 1951).
Varignana	*Cronaca B detta volgarmente Varignana*, ed. Albano Sorbelli (RIS², 18:1:2; Città di Castello, 1911).
Vauchez	André Vauchez, 'Une campagne de pacification en Lombardie autour de 1233', *EFRMAH* 78 (1966), 519–49.
Ver. 1228 Stat.	*Liber Juris Civilis Urbis Veronae*, ed. Bartolomeo Campagnola (Verona, 1728).
Ver. 1276 Stat.	*Gli statuti veronesi del 1276 colle correzioni e le aggiunte fino al 1323*, i, ed. Gino Sandri (Venice, 1940).

Vercelli Stat.	*Statuta Communis Vercellarum ab Anno MCCXLI*, ed. Giovambatista Adriani (Turin, 1876).
Verci	Giovanni Battista Verci, *Storia della Marca trevigiana e veronese*, (20 vols.; Venice, 1786–91).
VF	*Vitae Fratrum Ordinis Praedicatorum*, ed. Benedictus Maria Reichert (MOFPH 1; Louvain, 1896).
Vic. Stat.	*Statuti del comune di Vicenza 1265*, ed. Fedele Lampertico (Venice, 1886).
Villola	Pietro and Floriano da Villola, *Cronaca*, ed. Albano Sorbelli (RIS², 18:1:2; Città di Castello, 1911).
Vita Norberti B	*Vita Sancti Norberti Archiepiscopi Magdeburgensis [B]*, ed. Daniel Papebroch (AS 21: Jun. I, 826–7).

Introduction

THE period from the late 1100s to the mid-1200s was crucial in the formation of the political life of later medieval Italy. During this period Italian cities developed bureaucratic governments with wide authority and ability to intervene in the life of their citizens.[1] But the more orderly centralization of power that became normal in the Italian cities after the 1250s was by no means characteristic of the first half of the century. As a result of popular revolutions during the first two decades of the thirteenth century, Italian cities became progressively marked by the dual bane of an over-abundance of governing bodies and a dearth of legitimacy. The religious revivals that are the subject of this book occurred in a world where numerous individuals and organizations fought to exercise public authority, keep the peace, and defend themselves against enemies, but very few of them had any universally recognized right to do so. In such a world, political power followed directly on the popular perception of one's right to exercise it. The preachers who sparked the revival known as the Great Devotion of 1233, and other devotions like it, exercised, almost routinely, vast political authority over the cities in which they preached. They rewrote laws, deposed officials, and ordered the peace. This book is a study of how these mendicant revivalists rose to such vast, if temporary, political power and the conditions, attitudes, and beliefs that allowed them to do it.

POLITICAL AND SOCIAL BACKGROUND

The background of thirteenth-century Italian revivalistic preaching was the political world of communal Italy; the revivalists did not create this world, and its most important features pre-dated their activity by several

---

[1] For background of this period in Italian history recourse may be made to the most recent general surveys, on both of which I depend: David P. Waley, *The Italian City-Republics* (New York, 1969), and J. K. Hyde, *Society and Politics in Medieval Italy: The Evolution of the Civic Life, 1000–1350*, (New Studies in Medieval History; London, 1973). These surveys now replace those classics of 19th-cent. scholarship, J. C. L. Sismondi, *History of the Italian Republics in the Middle Ages*, ed. William Boulting (London, 1906), and W. F. Butler, *The Lombard Communes* (New York, 1906).

decades.[2] From the death of the Emperor Henry V in 1125 until the arrival in Italy of Frederick Barbarossa in 1154, imperial government effectively ceased in Italy, and the Italian cities appropriated to themselves the jurisdiction previously exercised by the emperor or his vicars. In 1158, at the diet of Roncaglia, the Emperor Frederick asked the leading lawyers of the cities of north Italy to define for him the powers and rights of the emperor over the cities of the Kingdom of Italy, that is, over what had once been the lands of the Lombards in central and northern Italy. These experts, steeped in the absolutism of Justinian's Corpus, sketched for him an exalted vision of imperial authority that, if instituted, would have reversed not only the recent usurpations of authority by the north Italian cities but the ecclesiastical liberties won by the Gregorian Reform as well. The imperialist lawyers became widely unpopular among their fellow citizens. The emperor, after years of bitter struggle to reassert his rights against the cities, which had united against him in 1176 to form the so-called First Lombard League, abandoned this ambitious political programme at the Peace of Constance in 1183.[3] This *de facto* imperial recognition of municipal self-government set the stage for the regime of independent city-states that ruled central and northern Italy in the high and late middle ages.[4] Imperialist sympathies lingered on among the lawyers of Bologna but even there the imperial cause became unpopular after Emperor Frederick II imposed the ban of the empire on the university for its adherence to the Second Lombard League, which had been formed to check his attempt to revive the old imperial rights in the north.

By the Peace of Constance, the emperor granted the cities of the Lombard League the exercise of both the imperial rights of jurisdiction, known as the *regalia*, and the often overlapping local rights of jurisdiction traditionally exercised by them, known as the *consuetudines*. He approved

[2] The generally accepted view of the rise of the communes is that of Giocchino Volpe, succinctly expressed in his 'Liber Maiolichinus', *Archivo storico italiano*, 5th. ser., 37 (1906), 93–114, and elaborated in his *Medio evo italiano* (Florence, 1961). This work is supplemented and expanded by Gina Fasoli, 'Gouvernants et gouvernés dans les communes italiennes du XIe au XIIIe siècle', *Receuils de la Société Jean Bodin*, 25 (1965), 47–96, and Paolo Brezzi, *I comuni cittadini italiani: Origine e primitiva costituzione* (Milan, 1940).

[3] Sources for the League are collected in Cesare Vignati, *Storia diplomatica della Lega Lombarda* (Milan, 1867; repr. Turin, 1966).

[4] On the reduced imperial role in early 13th-cent. north Italy, see Paolo Brezzi, 'I comuni cittadini italiani e l'Impero medioevale', *Nuove questioni di storia medioevale* (Milan, 1969), 177–207. For the Emperor Frederick II, David Abulafia's *Frederick II: A Medieval Emperor* (London, 1988) now provides an alternative to Ernst Kantorowicz's controversial *Frederick the Second, 1194–1250* trans. E. O. Lorimer (New York, 1957).

all of the latter, whether exercised inside or outside the city walls, even if the city could show no special grant of the right. This regularized municipal exercise of *regalia* and *consuetudines*, and recognized the cities' control of family law, commercial law, inheritance, land tenure, public order, and the making of new laws. About the only imperial prerogative that remained intact was minting, but even that could be exercised by a city after imperial authorization. Lawyers unaffected by imperial sympathies gave the Peace of Constance the broadest interpretation possible, and it came to be regarded as having approved in advance almost any measure that the cities might institute.

By the mid-1100s imperial rule in the north Italian cities had already been replaced by that of associations, sometimes known as *compagne*. Little evidence exists for the form taken by these governments in the inland cities, but comparison of what is known about Bologna with the regimes of Pisa and Genoa suggests that the twelfth-century communes all took similar forms.[5] The *compagna* was a temporary association of the citizens, subject to periodic renewal and under the direction of a group of consuls. With time it ceased to be temporary in all but theory, but it always seemed, more or less, a body existing outside of, and to some extent replacing, the older legal institutions of the imperial government. The number of consuls seems itself to have been subject to intermittent review. The consuls were elected by the commune and viewed as its servants. They could not make war or impose taxation without consent of the majority of those who made up the council of the commune. Their principal responsibilities were keeping the peace and administering justice. As early as 1115, there existed in Bologna an entity known as the *popolo* or people. It is likely that this was a formal sworn association similar to those known elsewhere as *compagne*.

In the course of the twelfth century these sworn associations had become known as 'communes'. It is safe to say that every major city of north and central Italy was ruled by a commune in the mid-1140s. These bodies were personal associations for the purpose of keeping the peace and administering justice. They have been described as the assembly of the citizens organized for some collective enterprise, such as self-government or war.[6] The mass of those who had taken the oath of membership composed an assembly, called in this period the *concio, arenga,* or

[5] On the rise of the commune in Genoa, see Franco Niccolai, *Contributo allo studio dei più antichi brevi della compagna genovese* (Milan, 1939).
[6] Hyde, *Society and Politics in Medieval Italy*, 54.

parlamentum. This body's involvement in government was limited to popular acclamation of the consuls' decisions. In practice, control of the commune belonged to the small group of families that dominated the consulate. The consuls were responsible to a council, consisting usually of about one hundred members. The revolutionary quality of this system lay not in its involving a large proportion of the populace in city government but in the fact that its authority was derived from the citizenry, not from a grant from above by the emperor or pope. In the eyes of academic lawyers and the emperor, such a body as the commune was highly irregular, unknown to Roman Law, and probably illicit. The commune was a concession, to be, at best, merely tolerated.[7]

As a public authority, the commune faced a rival in the local bishop, who had usually been the last agent to act in the name of the imperial government. Before the rise of the communes, the bishops of many cities of northern Italy had absorbed the old imperial office of count and ruled the city as representatives of the emperor. Such a fusion of secular and ecclesiastical roles became increasingly outdated after the Gregorian Reform, but the bishops retained far-reaching influence over secular affairs, even after they lost their secular office. They still exercised rights of jurisdiction over lands pertaining to their bishopric and in times of tension could become the symbol of unity in a city. Commune and bishop had usually reached a *modus vivendi* by the mid-1100s, and co-operation between ecclesiastical and civil authorities remained the rule until the consular regime was upset by the popular revolutions of the early 1200s. After that, the relationship between the legal rights of the bishop and the commune became the great cause of contention in city government.[8] After its governing position had been established, the commune underwent a rapid expansion. Routine judicial business was handled by panels of judges, leaving only appeals for the consuls to adjudicate. Treasurers appeared, who took over the administration of finances and property. Everywhere there proliferated a class of legal experts, the notaries, who, with eminent laymen, made up the seemingly endless parade of committees and councils to which administrative responsibilities were delegated.

The commune could, to a great extent, fill the vacuum left by the demise of the imperial administration, but it could do little to stem the

[7] Ibid. 64.

[8] There is no good general study of the relationship between the Church and the communes, but see the essays in *Vescovi e diocesi in Italia nel medioevo secoli IX–XIII* (Atti del II Convegno di Storia della Chiesa in Italia, Italia Sacra 5; Padua, 1964).

forces of corruption and violence which threatened the order of the city and perverted the administration of justice.[9] If anything, the proliferation of offices made this danger even more pressing. With good reason, the early communes were suspicious of unchecked power. The office of consul was temporary, and rotation in office was the rule. While such an arrangement did serve to check the ambitions of contending parties within the commune, it did not make for effective government. Already, in the mid-1100s, the communes had begun to experiment with a single executive, who would eventually come to be known as the *podestà*.[10] By the year 1151 there existed a *rector civitatis* at Verona.[11] In Bologna, a certain Guido of Sasso held the title of *rector* or *podestà*—there is good reason to believe that he was the first to use the later title.[12] Local conditions dictated the nature of the single executive when it was instituted, but all *podestà* seem to have shared certain characteristics. Like Guido, who was a student of the great Romanist Irnerius, many had legal training. To ensure military experience, they were usually chosen from a noble family. The majority of the early *podestà* came from landed families of the lesser feudal nobility, often from the region between Bologna and Cremona in central communal Italy. They were, above all, foreigners to the cities in which they served, expected to be above the partisan politics of the city they governed and not beholden to any faction there.

Guido of Sasso served in Bologna for at least 4 consecutive years, but it eventually became common to limit the *podestà*'s term to a year, or even 6 months. By the 1210s, the *podestà* was a permanent feature of government in all the Italian communes except Rome and Venice. The *podestà* was the head of the commune, the guardian of the law, and the highest judge. He controlled the bureaucracy and was the chair of the various municipal councils. In war, he was the commander-in-chief of the urban militia. To fulfil these capacities, the *podestà* needed a large household of

[9] To the extent that violence in the communes stimulated the development of new political institutions, it may not have been a wholly destructive force. See Lauro Martines, 'Political Conflict in the Italian City States', *Government and Opposition*, 3 (1968), 69–91, and the essays in id. (ed.), *Violence and Civil Disorder in Italian Cities, 1200–1500* (Berkeley, Calif., 1972).

[10] On this office, see Vittorio Franchini, *Saggio di ricerche sull'istituto del podestà nei comuni medievali* (Bologna, 1912), and G. Hanauer, 'Das Berufspodestat im 13. Jahrhundert', *Mitteilungen des Instituts für österreichishe Geschichtsforschung*, 23 (1902), 378–426.

[11] On the origins of the Veronese commune, see L. Simeoni, 'Le origini del comune di Verona', *Nuovo archivio Veneto*, 25 (1913), 49–143.

[12] On Guido, see G. Rabotti, 'Contributo alla storia dei podestà prefedericani', *Revista di storia del diritto italiano*, (32), 249–66, and Hyde, *Society and Politics in Medieval Italy*, 100–1.

experts in both legal and military matters. Men who had proved their
effectiveness as *podestà* in one city were often hired by another, and they
brought their advisers along with them. Each term of service was contracted between the new *podestà* and the commune. Since the appointing
of a *podestà* had often been triggered initially by factional violence or civil
war, the purpose of the institution was to provide a strong impartial
executive. In the late 1100s, at least, this seems to have been achieved.

The *podestà*-led government soon faced strong competition for domi-
nance in the commune. It was not the only sworn association that
developed to fill the vacuum left by the demise of the imperial administration, and its claim to public authority was by no means exclusive.
The increasing anarchy of the late 1100s urged the creation of other
sworn associations dedicated to the defence of their members. Powerful
families established *consorterie*, tower-societies to provide for the construction of a fortified stronghold in the city and for the collective
administration of estates in the countryside.[13] Such bodies provided
protection for their members during feuds and private wars and prevented
the fragmentation of property. For their members they served functions
similar to those of the commune—peace-keeping and protection. It is
possible that such organizations antedate the commune, which may itself
have been developed as a mechanism for curbing private wars between the
consorterie.

Within the city walls other associations proliferated.[14] By 1154
Piacenza had, along with its communal consuls, a consul of the merchants, who represented an association of merchants. The first evidence of
a merchant guild in Bologna dates to 1194.[15] By the early 1200s, there
existed at Bologna a group called the Società delle Arti, organized along
craft lines.[16] These bodies, like the noble *consorterie*, provided for peace-
keeping among their members and for their defence. Along with the
Arti, Bologna developed a group of Società delle Armi, which seem to

---

[13] On these associations of nobles, see Franco Niccolai, 'I consorzi nobiliari ed il comune
nell'alta e media Italia', *Revista di storia del diritto italiano*, 13 (1940), 116–47, 292–341,
397–477, and, more generally, F. Cusin, 'Per la storia del castello medievale', *Revista storica
italiana*, 7th. ser., 20 (1887), 25–58, 178–204.

[14] On these voluntary associations and their role in city life, see Gennaro M. Monti, *Le
confraternite dell'alta e media Italia* (2 vols.; Venice, 1927), and Franco Valsecchi, *Comune e
corporazione nel medio evo italiano* (Milan, 1949).

[15] Savioli, i. 2. n. 302.

[16] Associations of the *arti* begin to appear in the late 1100s in other cities as well. See
Pier S. Leicht, 'La corporazione italiana delle arti nelle sue origini e nel primo periodo del
comune', *Scritti vari di storia del diritto* (Milan, 1943), i. 297–308, 431–48.

have originated as neighborhood peace-keeping forces. Similar bodies existed by this time in the other north Italian cities. They, like the noble *consorterie*, became more and more 'communes in miniature', almost alternative governments. Public and private concerns became progressively intertwined. These associations provided protection, kept the peace, and provided for those in need. It was inevitable that such bodies would become a threat to the communal government under the *podestà*.

By the early 1200s noble *consorterie*, guilds, and protective associations had written statutes, oaths of membership, and elected leadership. In some cities these organizations existed peacefully alongside the communal government. In other cities, such as Belluno, the noble clans managed to dominate the commune completely. Where this did not occur, lawyers and wealthy merchants became the dominant force in the government. By the second decade of the thirteenth century, the membership of neighbourhood associations and guilds, along with deserters from the traditional governing class, began to organize, their organization usually taking the name *popolo*, the people.[17] Beginning in 1207 at Vicenza, the *popolo* of the north Italian cities began to move towards taking over the communal governments themselves. The Vicenzan chronicler Gerardo Maurisio describes the *popolo* as a kind of second commune, and his label is apt.[18] The members of the *popolo* remained members of the other organizations of the city and in Bologna, where these organizations were called the Arti and the Armi, it was those groups themselves that joined to create the *popolo*.[19] The Bolognese *popolo* was a vehicle of political involvement for the less well-to-do tradesmen, the foot soldiers of the urban militia, and transplants from the countryside, all of whom had established themselves in the city but had not broken into the traditional circles of power. It also included those rural landholders who remained outsiders to the knightly class. This new organization provided yet another contender for control of the city. In the eyes of legists, the *popolo*'s legal status was even more dubious than that of the commune. During the 1220s and 1230s, in city

[17] On the rise of the *popolo*, see Giovanni de Vergottini, *Arti e popolo nella prima metà del secolo XIII* (Milan, 1943); on their statutes and government, see id., 'Note sulla formazione degli statuti del popolo', *Revista di storia del diritto italiano*, 16 (1943), 61–70.

[18] Maurisio, 11; see also Giovanni de Vergottini, 'Il "popolo" di Vicenza nella Cronaca ezzeliniana di Gerardo Maurisio', *Studi Senesi*, 48 (1934), 354–74.

[19] The Bolognese *popolo* is well known from the studies of Gina Fasoli, 'Le compagnie delle armi a Bologna', *L'Archiginnasio*, 28 (1933), 158–83; 323–40, and ead., 'Le compagnie delle arti a Bologna fino al principio del secolo XV', *L'Archiginnasio*, 30 (1935), 237–80; their statutes are edited in Augusto Gaudenzi, 'Gli statuti delle Società delle Armi del Popolo di Bologna', *Bollettino dell'Istituto Storico Italiano*, 8 (1889), 7–74.

after city, the *popolo* demanded a direct role in government and, often after a violent revolution, achieved it. By 1250, the *popolo* was constitutionally recognized as the dominant political entity in all the major communes.

Leadership of the *popolo* varied from city to city, but certain common traits are visible. The early *popolo* usually had one dynamic leader who represented a group of officials from the neighbourhood organizations and guilds. The composition of these organizations is often difficult to distinguish. At Bologna these officials were known as the 'ancients' and were elected by the Armi and the Arti. Eventually the ancients became themselves a fixture of city government, serving a fixed term and living a common life, cloistered from the world like the *podestà*, during their term of office. The *popolo*'s development parallels that of the commune. By 1255 Bologna had a 'Captain of the People' to parallel the *podestà* of the commune. In the first half of the thirteenth century, however, the *popolo* had not yet developed a fixed single executive. It added a new element of instability to city government, undermining the communal executive but not replacing it.

Such conditions only exacerbated the tendency towards faction and internal violence. The often well-founded suspicion of other factions and individuals encouraged families and organizations to trust in self-help, both for mediation of their disputes and protection against enemies. It was not uncommon for two or three webs of alliances (to call them 'parties' would be to suggest a degree of organization that they did not have) to develop within a city. Civil war could break out at any time, and victory by one group could result in a massive departure of exiles.[20] Every city had a body of exiled citizens, often congregated in a traditionally hostile nearby city, waiting to return and eject those who had ejected them. Inter-city alliances grew up, often based on nothing more than the principle that my neighbour is my enemy, my enemy's neighbour is my friend. Historians have discerned two networks of inter-communal alliances in early thirteenth-century north Italy: one more or less pro-imperial, drawing support from the rural feudal nobility, and another, more anti-imperial than pro-papal, that had found political expression as the Second Lombard League. In the first decades of the thirteenth century the papacy was seldom able to intervene directly in north Italian politics; its involvement was occasional and opportunistic. Imperial authority was of little account. Emperor Frederick II was first a minor, and on reaching

---

[20] On exile and its role in Italian city life, see Randolph Starn, *Contrary Commonwealth: The Theme of Exile in Medieval and Renaissance Italy* (Berkeley, Calif., 1982).

majority he became involved in a debilitating struggle with the papacy. Frederick's interventions in north Italy during the 1220s were fruitless, and it was only after 1235 that he pieced together a network of alliances in the north that allowed him to control some of the region's communes through imperial vicars. Before the reaction against imperial successes that followed in 1237, there was little ideology underlying these two alliances—opportunism often dictated affiliation. The network of alliances did, however, guarantee political exiles a welcome in nearby towns, from which they could plan the overthrow of their own city's ruling party and their return.

Inside the communes, the creation of rival organizations and the rise of the *popolo* created a political crisis. The communal government lost its always-fragile and never-perfect monopoly over the use of force. The communal government found itself pushed aside more and more by bodies whose legitimacy was even more dubious than its own. And governmental authority was more and more diffused in the proliferation of offices and organizations. Early thirteenth-century communes suffered from too much government and too little legitimacy. At the same time the traditional understanding between the Church and the commune was upset, and jurisdictional conflicts multiplied. Much more can be said about the political life of individual north Italian cities during the late 1100s and early 1200s, but this must wait until later chapters, where the religious and political peculiarities of those cities will be described in greater detail as background to the revivals of 1233.

REVIVAL PREACHING IN COMMUNAL ITALY

It is against this general backdrop of political confusion, social upheaval, and urban violence, that one must place the religious life of the thirteenth-century commune. Social groups allied with the *popolo* turned more and more to new religious movements like the mendicants to provide for their religious needs. The bishop and his clergy looked suspiciously like the allies of the old regime. This religious life often took on enthusiastic, or even markedly revivalistic, qualities.[21] When religious enthusiasm

---

[21] On revivalism as a modern social and religious phenomenon, see William G. McLoughlin, jr., *Modern Revivalism: Charles Grandison Finney to Billy Graham* (New York, 1959) and id., *Revivals, Awakenings, and Reform: An Essay on Religion and Social Change in America, 1607–1977* (Chicago, 1978).

surfaces in our sources, it often centres on an itinerant preacher, commonly from a mendicant order.[22] His most visible activity was the public sermon, delivered to mass audiences, and reportedly marked by miraculous healings and divine interventions. Typically too, this preacher's activity in the commune did not remain merely religious but had political overtones. There is a tendency, especially among academics and intellectuals, to see such enthusiastic religion (and its impact on political life) as 'archaic', or as restricted to the unsophisticated and ill-educated. This prejudice, and the desire to distance oneself from such behaviour, is not new. One nineteenth-century Englishman ascribed the revivals of 1233 to the 'Italian temperament', with its lack of 'any great depth of moral earnestness'.[23] The Swiss Carl Sutter, who in 1891 wrote the last full-scale history of the Great Devotion, relegated the analysis of its revivalism to social psychologists and psychiatrists[24]—for him the political successes of the mendicant demagogues and the religious exaltation of 1233 were merely the results of 'nervous over-excitement'.[25] For certain observers the temptation to relegate emotionalistic religion to the realm of the 'inferior races' and the mentally disturbed has always been strong.

Whatever one thinks of the phenomenon, medieval Italy saw varying receptivity to revivalistic religion. Some revivals seem spontaneous lay-movements without clerical leadership.[26] In the north Italian cities of the

---

[22] A fine, concise introduction to preaching in communal Italy may be found in Carlo Delcorno's *La predicazione nell'età comunale* (Florence, 1974), which contains bibliography and representative texts.

[23] J. A. Symonds, 'Religious Revivals in Medieval Italy', *The Cornhill Magazine*, 31 (1875), 54; cited with enthusiasm by G. G. Coulton, *From St. Francis to Dante: Translations from the Chronicle of the Franciscan Salimbene (1221–1288) with Notes and Illustrations from other Medieval Sources*, 2nd. edn. (1907; repr. Philadelphia, 1972), 21.

[24] Sutter, 170: 'Sie nach allen Seiten zu untersuchen und zu erklären, wäre eine Aufgabe für den Sozialpsychologen, man möchte fast sagen für den Psychiater.' This passage is omitted in the Italian translation.

[25] Sutter, 170: 'nervöse Überreizung'.

[26] One Italian revival often seen as a spontaneous lay movement is that of the Flagellants in 1260. Since the focus of this book is revivalistic preaching, not revivalism as such, the reader seeking more on 1260 is referred to the collection of essays in *Il movimento dei Disciplinati nel settimo centenario dal suo inizio (Perugia-1260)* (Convegno Internazionale, Perugia, 25–28 settembre 1960; Perugia, 1962). Raffaello Morghen, 'Ranieri Fasani e il movimento dei Disciplinati del 1260', ibid. 38–9, draws a strong contrast between 1230 and 1260, noting the lack of flagellation and vicarious expiation in the former, and the lack of overt political involvement in the latter; Gilles-Gérard Meersseman, 'Disciplinati e penitenti nel duecento', ibid. 49, also contrasts the lay leadership of 1260 with the clerical leaders of 1230. I second and emphasize these contrasts. 1260 was a rival of great importance but one that does not fit into the tradition that is the subject of this study. If

early thirteenth century the catalyst was often a preacher. This centrality of preaching in popular piety is reflected in the period's ideas of sanctity; the most popular saints of the thirteenth to fifteenth centuries were almost all preachers.[27] Some even saw the rise of mendicant preaching as ushering in a New Age. One Dominican of the mid-thirteenth century, Ambrose Sansedoni, had occasion to speak on the ages of the Church. He described with apocalyptic, almost ecstatic, exhilaration how there had earlier been the prophets who heralded the dispensation of the Old Testament, the age of the Law. Then came the greatest prophet of all, Jesus Christ, the initiator of the Age of Grace. His new dispensation had its own ministers. First the Apostles, and then the bishops, had proclaimed the Gospel. In the last days, when the hour of the bridal banquet was drawing near, God had sent his servant, the preacher Dominic, to announce with greater vigour the Good News of God. Dominic had ushered in a New Age, no longer that of the prophets, Apostles, or bishops, but that of the preachers. Now in the final age, the Friars Minor and Friars Preachers were carrying on Dominic's work.[28] In the 1260s when Ambrose preached, his vision of three ages may have owed something to Joachim of Fiore, who had spoken of two preachers who would lead in the last age, but his enthusiasm for preachers was found outside apocalyptic circles as well.

Preachers continued to spark outbursts of religious enthusiasm throughout the later middle ages and use that religious enthusiasm as an instrument for political action. Fourteenth-century Italy saw the austere James of Bussolaro exercise an energetic preaching ministry in Pavia during the 1340s and 1350s, and play an important role in the politics of

---

there is an element of continuity between 1233 and 1260 it is the element of peace-making, as noted by Hermann Hefele, *Die Bettelorden und das religiöse Volksleben Ober- und Mittelitaliens im XIII. Jahrhundert* (Leipzig, 1910), 131–2, and by Gary Dickson, 'The Flagellants of 1260 and the Crusades', *Journal of Medieval History*, 15 (1989), 237. The later study is broad, suggestive, and with *Il movimento dei Disciplinati*, provides a suitable introduction to the state of studies on 1260.

[27] The anthropologist Ida Magli, *Gli uomini della penitenza: Lineamenti antropologici del medioevo italiano* (Milan?, 1967), 129, gives as examples Francis, Dominic, Bernardino of Feltre, James della Marca, Anthony of Padua, John of Capistrano, and Vincent Ferrer.

[28] Ambrose Sansedoni, *Sermones de Tempore*, Siena, Biblioteca Comunale MS T. IV. 7, fo. 80^{r-v}: 'Postea in novo testamento per apostolos et postea per episcopos, sed modo quia iam est hora cene, misit servum suum, id est, beatum Dominicum.' He uses the same image later (fo. 82^r), replacing St Dominic with 'ordinem minorum et predicatorum'. Ambrose's view was not unique. See also Jordan of Saxony, *Liber de Principiis Ordinis Praedicatorum*, ed. H.-C. Sheeben (MOFPH 16: MHD 2; Rome, 1935), 25, for whom the Dominicans served the Church of the last age.

the city until Galeazzo Visconti, who came to power there in 1359, gaoled him. In the fifteenth century, the Franciscan Bernardino of Siena electrified Tuscany, Lombardy, and the Marche by his preaching and promoted the cult of the Holy Name. Although he was a great political peace-maker, Bernardino generally remained aloof from government. The same century witnessed the rise of the Dominican Savonarola at Florence. He used his preaching to dominate Florentine politics until he met his martyrdom for attacking the Borgia pope, Alexander VI.

This book is not a comparative study of revivalists and politics in late medieval Italy. Rather, it focuses on what is probably the oldest example of such a combination of preaching and political activity, that which occurred during 1233 in the region now known as Emilia-Romagna and the Veneto. It is probably the clearest example of that widespread resurgence of popular preaching that marked the late twelfth and early thirteenth centuries.[29] This resurgence found its major practitioners in the Dominican and Franciscan orders. What distinguishes the events of 1233 from the rest of contemporary mendicant preaching is that it can be identified, and contemporaries identified it in itself, as a distinct religious event, a 'Devotion', what one would today call a 'revival'. Contemporary observers' characterization of the revival as a single phenomenon was no accident—its preachers worked in concert to promote religious enthusiasm and direct it to religious and political ends.

If one omits the earlier preaching-campaigns to promote the crusades, this Devotion is probably the first of the intermittent upsurges of popular religiosity sparked by preaching in the later Italian middle ages. Its most noticeable characteristic was the use of preaching as a springboard into direct political involvement and peace-making. One can trace in detail the rise and fall of only one of its preachers, the Dominican John of Vicenza, but there exists for the revival such abundant evidence that the student can, to a great extent, enter into the religious world of the Devotion and see how it was intertwined with political life. It is also possible to move out from the isolated campaign of 1233 to examine the activities and methods of other preachers whom contemporaries identified as practising the same form of revivalism in the years following the Great Devotion.

---

[29] A phenomenon admirably described by D. L. d'Avray, *The Preaching of the Friars* (Oxford, 1985), 25–9.

METHOD, STRUCTURE, AND SOURCES OF THIS STUDY

This study departs from one approach currently very popular in the study of medieval preaching. Recent attempts to reconstruct the realities of medieval preaching have been above all textual, focusing on the medieval sermon.[30] D. L. d'Avray's book *The Preaching of the Friars* exemplifies this trend. D'Avray's first goal is to describe medieval 'preaching aids'—sermon collections, manuals for preachers, etc.—and to show how they became a kind of international 'mass media'. Secondly, after identifying Paris as the centre of the mass media, he treats the means by which Paris stationers distributed homilies. Thirdly, he points out that thirteenth-century sermons, as found in extant texts, were not a popularization of the 'scholastic' way of thinking—dialectics and questions—but rather had a method and logic all their own. Finally, he proposes that sermons, because of their universal diffusion and acceptance, can give us insight into hitherto poorly understood medieval patterns of thought such as the tendency to cluster concepts in threefold groupings analogous to the famous triads *intellectus−affectus−effectus* and *fides−proles−sacramentum*. In comparison with the vast number of sermon collections and preaching-aids available in published and manuscript form, medieval descriptions of preachers and preaching are, according to d'Avray, rare, stereotyped, and fragmentary. For him, such descriptions are dangerously deceptive because they usually portray exceptional figures, such as Berthold of Regensburg (d'Avray's example), rather than day-to-day preachers.[31] This characterization of the sources lends theoretical support to the current emphasis on sermon texts.

Remarkable and enlightening as the results of recent sermon studies may be, one may wonder if it has not been a mistake to devalue so strongly narrative descriptions of medieval preaching. To restrict attention almost exclusively to medieval sermon-texts implies, perhaps, an overly intellectualized image of preaching. Current sermon-studies are yielding important results, but one must place these results in the context of the pulpit and the piazza. One student of modern revivalism has remarked that the sermon notes and printed texts left by the great American revivalists of the past hundred years are of little use for under-

---

[30] For recent scholarship, see Carlo Delcorno, 'Rassegna di studi sulla predicazione medievale e umanistica', *Lettere italiane*, 33 (1981), 235−76. On Italian medieval preaching see R. Rusconi, 'Predicatori e predicazione', in *Annali della storia d'Italia*, iv (Turin, 1981), 949−1035.

[31] D'Avray, *Preaching of the Friars*, 61.

standing the phenomena of revivalism, because the sermon is a spoken, not a written, word.[32] He could easily be describing medieval preaching as well. The experience of medieval preaching included what the hearers saw and felt, as well the words they heard, or often, more likely, failed to hear.[33] To retrieve the non-verbal elements of medieval preaching requires a re-evaluation of narrative (and legal) sources, because these are the only evidence for the non-textual elements of preaching.[34] Furthermore, these often supply us with fascinating insights into the medieval preacher's persona. The student of medieval revivalism now knows much about the history of the medieval sermon, but the history of preaching and medieval preachers remains to be written.

When modern scholars have categorized medieval preaching, as opposed to medieval sermons, they have done so by centuries, by religious orders,[35] or as 'university' and 'popular'.[36] Sermon studies by scholars like d'Avray demand that these categories be rethought. The student of medieval preaching would best begin with the identification of its varieties as they were distinguished by medieval writers themselves, not with its submersion into vast, perhaps a priori, frameworks. To write a complete history of the medieval preacher would be an immense undertaking, requiring a knowledge not only of medieval sermon-texts and the modern scholarship on them, but also of the other, usually narrative, descriptions of medieval preaching. It would also require a knowledge of 'heretical' preaching in all its forms. Such a project is, as yet, impossible. One might begin with a study of some traditional category of medieval preaching, such as 'itinerant preaching', 'mendicant preaching', 'heretical preaching', and so on, but this would risk imposing on a past age an anachronistic modern category. Even a categorization which might seem rooted in history—e.g. a contrast between 'intellectual' Dominican

[32] McLoughlin, *Modern Revivalism*, 531.

[33] Carlo Delcorno, 'Origini della predicazione francescana', in *Francesco d'Assisi e francescanesimo dal 1216 al 1226* (Atti del IV Convegno Internazionale, Assisi, 15–17 ottobre 1976; Assisi, 1977), 140–1, rightly emphasizes the non-verbal effects of preaching, noting that although those who heard homilies of St Francis of Assisi were deeply moved, they could not later remember what he had said.

[34] Carlo Delcorno, 'Predicazione volgare e volgarizzamenti', *MEFR* 89 (1977), 681, suggests this method for the study of early medieval preaching, since sermons are lacking.

[35] This may be a fading trend: Louis-Jacques Bataillon, 'La predicazione dei religiosi mendicanti del secolo XIII nell'Italia centrale', *MEFR* 89 (1977), 694, shows that contrary to popular opinion, no distinction exists between an 'intellectual' Dominican style of preaching and an 'emotional' or 'popular' Franciscan style. The older view is taken by Angelus Maria Walz, *Compendium Historiae Ordinis Praedicatorum* (Rome, 1930), 159, who suggests that early 13th-cent. Dominican preaching was 'spontanea, ardens' and that it later became more formal.

[36] Delcorno's *La predicazione nell'età comunale*, for example, uses this division.

preaching and 'emotional' Franciscan preaching—can, as this study will show, be very deceptive.[37]

To avoid stereotype and anachronism, this study will begin with a revival that occurred during 1233 in Emilia-Romagna and Lombardy. Salimbene de Adam of Parma, its chief chronicler, called it 'The Devotion of the Alleluia', and this is the name by which it was commonly known. It touched nearly every city of north Italy. Some have distinguished it from a nearly simultaneous 'Great Devotion of the Friars Preachers',[38] but Salimbene's language is probably not precise enough to allow such a distinction. In any case, this revival has a special significance in the history of medieval preaching, because Salimbene identifies a *distinctive* homiletic style as typical of its preachers. For want of a better name, he called this style 'the way of preaching of the ancient preachers as they preached at the time of the Alleluia'.[39] It is strange that Salimbene's identification of this preaching as peculiar in style has not received greater notice.[40] Salimbene presents the Alleluia as only one strand in a varied and rich tapestry of medieval Italian preaching. The purpose of this book is to reconstruct that style of preaching. Since these revivalists of the 'Ancient Way' were always politically active as peace-makers and some-times as legislators, these activities must also be integrated into the analysis of this form of revivalism.

Salimbene himself tells us that while the revivalism practised by the preachers of 1233 was unusual, it was not unique to that year. Salimbene spoke of his friend Barnabas of Reggio, who died in 1285, as knowing 'how to represent the way of preaching of the ancient ('antiquorum') preachers, the way in which they preached at the time of the Alleluia, when they put themselves forward as working miracles'.[41] Three aspects

---

[37] Or, at least, that seems to be the case for the early and mid-13th cent. Daniel R. Lesnick, *Preaching in Medieval Florence: The Social World of Franciscan and Dominican Spirituality* (Athens, Ga., 1989), would uphold the traditional scholastic/exhortative distinction between Dominican and Franciscan preaching in Florence during the late 13th cent. While his sociological analysis of the clientele for the two orders does suggest reasons for such a division, one can only wish that he had presented evidence from a wider number of sermons to support its actual existence.

[38] For example, see Fumagalli, 257–72.

[39] Salimbene, 595: 'modum predicandi antiquorum predicatorum, secundum quod predicabant tempore Alleluie'.

[40] Mariano d'Alatri, 'Predicazione e predicatori francescani nella *Cronica* di fra' Salimbene', *Collectanea franciscana*, 46 (1976), 87, for example, sweepingly concludes that for Salimbene 'la predicazione dei francescani nel Duecento fu di carattere eminentemente parenetico-morale ed escatologico'.

[41] 'Et sciebat representare modum predicandi antiquorum predicatorum, secundum quod predicabant tempore Alleluie, quando intromittebant se de miraculis faciendis, ut diebus illis oculis meis vidi.'

of the Ancient Way stand out in this text: the revivalism typical of the Alleluia was practised by other preachers (or by Barnabas, at least), by 1285 Salimbene considered it a dead art ('antiquorum'), and it involved miracle-working. Because Salimbene speaks of this style as flourishing between 1230 and 1280, descriptions of similar mendicant preaching in Italy before the 1280s will be freely drawn upon, if the preachers conform, to a lesser or greater extent, with what was typical of the revivalists of 1233.[42] The miraculous element will have to be examined in greater detail later.

This study begins with what is best known: the actual preachers of the revival. To find the core group of the revival of 1233 was simple. Salimbene himself numbered several acquaintances among the revival's 'Solemn Preachers'.[43] One, Gerard of Modena, had interceded with Brother Elias for Salimbene's admission into the Franciscan order.[44] Near the end of his chronicle, on the occasion of the obituary of the same Barnabas, Salimbene listed the major revivalists by name. They were John of Vicenza, Bartholomew of Breganza, and Jacopino of Parma, all Dominicans; as well as Gerard of Modena, and Leo, later archbishop of Milan, both Franciscans.[45] All, or at least all but Leo, he describes as known miracle-workers.

The Franciscans listed as participants in the Alleluia passed into the devotions and folklore of their order as blesseds; they were among the most important Italian Minorites of the first generation. The only Dominican on his list to be raised to the altar was Bartholomew of

[42] On revivalism and popular religion, see Raoul Manselli, *La Religion populaire au moyen âge: Problèmes de méthode et d'histoire* (Montreal, 1975), esp. p. 110, where he alludes to 1233.

[43] This phrase has generated some confusion. Gustavo Cantini, 'L'apostolato dei beati Gherardo Boccabadati e Leone Valvassori da Perego francescani e la devozione dell'Alleluia', *SF*, 3rd ser., 9 (1938), 349, for example, thinks that Salimbene used it only for high prelates. In fact he uses it to describe all the preachers of the Alleluia, as well as several others, not all of whom were prelates. As is obvious, if not from Salimbene, at least from *ACPR*, 11, the word refers to those who preached at *solemnes praedicationes*, that is to say, large formal preaching campaigns directed towards the general public, usually in large cities such as Florence or Rome.

[44] Salimbene, 75, writes: 'Hic pro me rogavit fratrem Helyam generalem ministrum ordinis fratrum Minorum, ut ad ordinem me reciperet.'

[45] Salimbene, 595: 'Isti fuerunt frater Iohannes de Vincentia, qui Bononie miracula faciebat et de ordine Predicatorum erat. Item frater Bartholomeus similiter de Vincentia de ordine supradicto, qui Parme miracula faciebat. Frater Iacobinus de Parma, qui in Regio miracula faciebat et ideo de Regio dicebatur, et de ordine fratrum Predicatorum erat. Item frater Ghirardus de Mutina ex ordine fratrum Minorum, qui totam Italiam circuibat predicando et miracula faciendo. Frater Leo archiepiscopus Mediolanensis ex ordine Minorum similiter, qui in Mediolano optime predicabat. Et multi alii, quos vidi et cognovi, quorum memoria sit cum Deo! Amen.'

Breganza, known to Salimbene as 'of Vicenza' because he was later bishop of that city.[46] For these sainted preachers, hagiographic descriptions are ample. For the others, even the famed John of Vicenza, one must glean disparate and fragmentary reports from secular chroniclers and mendicant story-tellers. From this material there arise the basic outlines of the Ancient Way. Once these are known, the investigation can widen and draw on the collections of 'family traditions' produced by thirteenth-century and fourteenth-century mendicant compilers. One must carefully discern the extent to which any such stories reflect actual events or have been stylized and stereotyped.[47] Descriptions of the preachers must be treated individually and with care and contrasted with those of other mendicant preachers of the period.

When a characteristic element of the Ancient Way has been identified with certainty, one can feel free to illustrate it with the examples found for similar preachers of the period. For this purpose, several personalities came instantly to mind and have proved very useful in this study. The Dominican Peter of Verona, today better known as Peter Martyr, and the Franciscan Anthony of Padua were immediate candidates. Anthony proved less useful since he spurned one already-noted characteristic of Alleluia preaching, miracle-working. The stories about Peter proved much more helpful.[48] Looking further afield, this work draws on the vita and stories told of the Dominican peace-maker and miracle-worker, Ambrose Sansedoni of Siena. He can be called without reservation a revivalist in the 'ancient style'. The preacher of the Fourth Crusade, Foulques of Neuilly, although a bit early and not Italian, also provides illuminating contrasts.[49] Along with other less-known preachers, these men will provide most of the outside witness to the style of revivalism

---

[46] Sources for the Devotion never describe Bartholomew's activity. He left a book of homilies, but these are literary creations dating from the time of his episcopacy. See Thomas Kaeppeli, 'Der literarische Nachlaß des sel. Bartholomaeus von Vicenza O.P.', in *Mélanges Auguste Pelzer* (Université de Louvain, Recueil de Travaux d'Histoire et de Philologie, 3rd ser., 26; Louvain, 1947), 275–301.

[47] On the assimilation of hagiographic stories to accepted models of behaviour, see Michael Goodrich, 'The Politics of Canonization in the Thirteenth Century: Lay and Mendicant Saints', *Church History*, 44 (1975), 300.

[48] Peter's actual role in the Devotion is unknown. He was certainly not, as is often said, inquisitor of Lombardy from 1232 to 1233; see Antoine Dondaine, 'Saint-Pierre-Martyr: Études', *AFP* 23 (1953), 67–73.

[49] D. L. d'Avray, *Preaching of the Friars*, 22–8, and A. Forni, 'La "Nouvelle Prédication" des disciples de Foulques de Neuilly: Intentions, techniques et réactions', in *Faire croire: Modalités de la diffusion et de la réception des messages religieux du XIIᵉ au XVᵉ siècle* (Collection de l'École Française de Rome, 51; Rome, 1981), 19–37, date the beginning of a new type of revivalistic preaching similar to that of 1233 to the late 11th cent. in general and to Foulques in particular.

here treated. The principle has been consistent: the character of the revival and the revivalists is to be determined from the descriptions of those who were known participants in it; descriptions of other preachers can then be drawn upon to cast light on a phenomenon when it has already been identified. As I focus on different aspects of the revival, I shall feel free to draw on some passages in the sources more than once, when these stories can throw light on more than one aspect.

I believe that this study is the first to treat the sociological, devotional, political, and legal aspects of thirteenth-century revivalistic preaching in a single work. This book is meant as a complement to the textual approach of contemporary sermon and preaching studies. It begins by focusing on the events of 1233, but goes beyond them to reconstruct the Ancient Way itself. These two projects are distinct, and each presents its own problems. In 1891, Carl Sutter, in his now classic study of 1233,[50] chose to give a chronological presentation of that revival and fit undated events and incidents into it at likely places. The result is a vigorous, if occasionally discursive, narrative, but it shows little appreciation of the dynamics of the revival itself. On the other hand, the events of 1233 are too important to be submerged and lost in the study of the Ancient Way as a religious phenomenon. To do so would deprive the reader of a direct entry into the event which was, for contemporaries like Salimbene, the chief exemplar of the Ancient Way itself. For these reasons this work falls into two parts, a narrative section recounting the events of 1233 and an analytical section drawing on all known practitioners of the Ancient Way.

Part I of this book consists of three narrative chapters, the first tracing the development of the Devotion of 1233 in Emilia, and the second and third its arrival in Bologna and extension north into the Veneto. Although access may now be had to a few texts unknown to Sutter in 1891, knowledge of events in Emilia during 1233 remains extremely fragmentary. The descriptions of the revivalists appear in chronicles, usually lay, for the cities of Piacenza, Parma, Reggio; for Parma we also have the narrative provided by Salimbene.[51] The Communal Statutes of

---

[50] *Johann von Vicenza und die italienische Friedensbewegung im Jahre 1233* (Freiburg, 1891), trans. Maria, Gelda and Olga da Schio, *Giovanni da Vicenza e l'Alleluja del 1233* (Vicenza, 1900).

[51] The chronicle sources for the revivals of 1233 in Emilia have been edited as follows. For Parma: *Annales Parmenses Maiores*, ed. Georg Heinrich Pertz (MGH.Ss. 18; Hanover, 1863), 790–9; *Chronicon Parmense ab Anno 1038 usque ad Annum 1338*, ed. Giuliano Bonazzi (RIS² 9:9; Città di Castello, 1902); *Chronicon Placentinum*, ed. Lodovico Antonio Muratori (RIS 16). For Piacenza: Muzio of Modena, *Annales Placentini Gibellini*, ed. Georg Heinrich Pertz, (MGH.Ss. 18; Hanover, 1863), 457–581; *Annales Placentini Guelfi*, ed. Georg

Parma also allow us to establish some dates for the events there. For Bologna there is, along with several other chronicles, important evidence from the chronicles of Girolamo de' Borselli, a fifteenth-century Bolognese Dominican.[52] Although late, he is usually trustworthy and may have had access to now lost materials in the Dominican Archives. The secular chroniclers who report the revival contribute varied and pregnant observations on the friars and their political projects. For the individual revivalists we can also draw on the collections of stories that compose the 'family histories' of the two mendicant orders.[53] Fragmentary as they are as narrative, the city chronicles do allow for a reconstruction of events in early 1233 and for establishing which revivalists were active in which cities.

Carl Sutter's history of these events, written in German and translated (with modifications) into Italian, has remained often inaccessible to English-speaking students of medieval revivalism. His finest work was his reconstruction of the preaching-campaigns of John of Vicenza. Chapters 2 and 3 here have followed his lead. Since access is now possible to documents unpublished in 1891, in particular the collection in the *Corpus Chronicorum Bononiensium*, one can now correct certain mistakes in

Heinrich Pertz (MGH.Ss. 18; Hanover, 1863), 411–57; Giovanni de' Mussi, *Chronicon Placentinum*, ed. Lodovico Antonio Muratori (RIS 16). For Modena: *Annales Veteres Mutinenses ab Anno MCXXXI usque ad MCCCXXXVI* (RIS 11, cols. 53–130). For Faenza: *Chronicon Faventinum Magistri Tolosani {AA 20 av. C.-1236}*, ed. Giuseppe Rossini (RIS² 28; Bologna, 1939). For Reggio: *Die Doppelchronik von Reggio und die Quellen Salimbenes*, ed. Alfred Wilhelm Dove (Leipzig, 1873); *Memoriale Potestatum Regiensium*, ed. Lodovico Antonio Muratori (RIS 8, cols. 1071–80).

[52] Girolamo Borselli, *Chronica Magistrorum Generalium Ordinis Fratrum Predicatorum*, Bologna, Biblioteca Universitaria MS 1999; the major printed chronicles are those in the Corpus Chronicorum Bononiensium, 2 ed. Albano Sorbelli (RIS² 18:1:2; Città di Castello, 1911); Girolamo Borselli, *Cronica Gestorum ac Factorum Memorabilium Civitatis Bononie*, ed. Albano Sorbelli (RIS² 23:2; Città di Castello, 1912); and Matteo Griffoni, *Memoriale Historicum de Rebus Bononiensium*, ed. Lodovico Frati and Albano Sorbelli (RIS² 8:2; Città di Castello, 1902).

[53] For the Dominicans the major collection of early stories is that commissioned by the Dominican master of the order Humbert of Romans in 1256, the *Vitae Fratrum Ordinis Praedicatorum*, ed. Benedictus Maria Reichert (MOFPH, 1; Louvain, 1896); and the collection by Thomas of Cantimpré, *Bonum Universale de Apibus*, ed. in part Joannes Baptista Sollerius (AS, 28: Jul. I; Paris: 1867), 424–5, commissioned by the same master of the order. For the Franciscans, besides the early lives of Francis and the *Fioretti*, the most useful collections are late: the anonymous *Chronica XXIV Generalium Ordinis Minorum* (AF 3; Quaracchi, 1894); Thomas of Pavia, *Dialogus de Gestis Sanctorum Fratrum Minorum*, ed. Ferdinand M. Delorme (Bibliotheca Franciscana Ascetica Medii Aevi, 5; Quaracchi, 1923); Bernard of Bessa, *Liber de Laudibus Beati Francisci* (AF 4; Quaracchi, 1912); and the very late Bartholomew of Pisa, the *De Conformitate Vitae Beati Francesci ad Vitam Domini Iesu* (AF 4–5, 2 vols.; Quaracchi, 1912–17).

his narrative of John's excursions to the Veneto. On several occasions, this book's reconstruction has departed from his chronology when other orderings seemed more compelling. For the developments in the Veneto we have two chronicle sources of prime importance: those of Gerardo Maurisio of Vicenza and Rolandino of Padua.[54] Both have been published in the new Rerum Italicarum Scriptores, and both take a negative view of Friar John. But these chroniclers, who pass over secular events of great importance in one or two lines, dedicated pages to the Alleluia preachers and their activities. These rich sources for how contemporaries viewed the revival have generally been overlooked.[55] There also exist a number of legal instruments and papal letters that allow for the dating of events in the North.[56]

Part II treats the revival and similar thirteenth-century preaching as a religious, social, and political phenomenon. A narrative of 1233 goes only part of the way towards explaining the Ancient Way of Preaching. This is partly because a narrative must ignore the many existing descriptions of the preachers that cannot be positioned in the chronology. Since these stories often carry us directly into the experiences of the hearers and participants, they provide perspective on the revival that a bare narrative cannot. The appearance and development of the revival in each individual city and John of Vicenza's campaigns as a whole show

[54] The most important contemporary chronicle sources for the Veneto campaigns are: the Paduan Rolandino, *Cronica in Factis et circa Facta Marchie Trivixiane*, ed. Antonio Bonardi (RIS² 8:1; Città di Castello, 1905); and the Vicenzan Gerardo Maurisio *Cronica Dominorum Ecelini et Alberici Fratrum de Romano*, ed. Giovanni Soranzo (RIS² 8:4; Città di Castello, 1914), which was abridged and supplemented in the 16th cent. by Antonio Godi, *Cronaca*, ed. Giovanni Soranzo (RIS² 8:1; Città di Castello, 1905). Gerardo Maurisio, who came from an old aristocratic family of Vicenza, wrote his chronicle in haste in 1237 to ingratiate himself with the victorious da Romano. It is a detailed if biased source for events in 1233. Other information concerning the northern campaigns can be gleaned from the *Chronicon Marchiae Trevisinae et Lombardiae*, ed. L. A. Botteghi (RIS² 8:3; Città di Castello, 1916); and the *Liber Regiminum Padue*, ed. Antonio Bonardi (RIS² 8:1; Città di Castello, 1905), 269–376. Especially important for Verona are Parisio of Cerea, *Annales Veronenses*, ed. Philipp Jaffé (MGH.Ss. 19; Hanover, 1866), 2–18; and the *Annales Veteres Veronenses*, ed. Carlo Cipolla, *Archivio Veneto*, 9 (1875), 77–98. These sources are analysed by Girolamo Arnaldi in his *Studi sui cronisti della Marca trevigiana nell'età di Ezzelino da Romano* (Rome, 1963).

[55] Lidia Capo, 'Cronache mendicanti e cronache cittadine', *MEFR* 89 (1977), 634, n. 2, continuing this trend, concludes that secular authors showed little interest in Alleluia preachers, except as 'maliziosi religiosi che fomentano le discordie'.

[56] Many documents relating to 1233 are edited in Giovanni Battista Verci, *Storia della Marca trevigiana e veronese*, i, app. (Venice, 1786); and, especially for John of Vicenza, in the *Bullarium Ordinis Fratrum Praedicatorum*, i, ed. Thomas Ripoll (Rome, 1729).

that the revival everywhere followed a similar pattern. The preachers first created an audience, then gave that audience the sense of collective identity, and finally used this power-base to launch a social and political programme. To understand the dynamism of the revival, one must first examine the means by which the revivalists attracted and united their audience and the way in which they presented themselves to it. Chapter 4, the first chapter of this book's second part, treats what can be known about this.

It has been noted that at one point in his chronicle, Salimbene made an interesting remark about the Ancient Way. In his view, miracle-working was so essential to it that, when speaking of the Dominican Bartholomew of Breganza and his founding of the Militia of Jesus Christ during the Flagellant Devotion of 1260, he distinguished that devotion from the other 'Great Devotion' of his time, 1233, by the role of the miraculous in the latter.[57] That a certain miraculous quality surrounded revivalists is not surprising, nor is the report that they were occasionally credited with healings or seemingly miraculous conversions. Such is not Salimbene's sense. His report, which other independent witnesses confirm, is that the preachers consciously and deliberately exploited their reputation for working miracles and that they made miracle-working an integral part of their preaching. Since miracle-working was of such importance in the creation of an audience and in giving a particular character to the revival, its special role in the gathering and uniting of an audience is the topic of Chapter 5.

That chapter will also draw on the images of the revivalists and the attitudes expressed towards them in what may loosely be called hagiographic sources.[58] These reports almost always take the form of a marvel

[57] Salimbene, 467; 'Item recordor, quod ordo iste [militum Jesu Christi] factus fuit in Parma tempore Alleluie, id est tempore alterius devotionis magne, quando cantabatur 'Alleluia', et intromittebant se fratres Minores et Predicatores de miraculis faciendis, anno Domini M°CC°XXX°III°, tempore pape Gregorii noni.'

[58] The usefulness of these sources as a source for popular and learned perceptions of holiness has been amply demonstrated by André Vauchez's magisterial *La Sainteté en occident aux derniers siècles du Moyen Âge d'après les procès de canonisation et les documents hagiographiques* (Rome, 1981). Since 'sanctity' does not seem to be the characteristic most commonly ascribed to the preachers under consideration, I shall draw on documents more 'folkloric' than hagiographic and treat the available hagiography in a way different from that of Vauchez. Saints' lives used in this study include: for Dominic, the *Acta Canonizationis Sancti Dominici*, ed. Angelus Walz (MOFPH 16: MHD 2; Rome, 1935), 91–194; for Anthony of Padua, the earliest life, the *Vita prima di s. Antonio o 'Assidua' (c.1232)*, ed. and trans. Vergilio Gamboso (Fonti agiografiche antoniane 1; Padua, 1981); for the Bl. Ambrose Sansedoni the life by his contemporary Gilsberto Alessandrino *et al.*, *Legenda Antica Beati Ambrosii Senensis*, ed. Daniel Papabroch (AS 8: Mar. III; Paris, 1865), 180–209; and for

or miracle story. Those revivalists who later became the focus of a
cult—e.g. Leo de' Valvassori, Peter Martyr, Bartholomew of Breganza,
Anthony of Padua—were sometimes the subject of saints' lives. But for
the most part I shall draw on story-collections produced as family history
by the mendicant orders. Certain principles have guided the use of these
stories. I have favoured stories that do not seem to have conformed to
traditional hagiographic stereotypes. Since this study is interested principally in the image of the Ancient Preachers, it gives pride of place to
stories told of those revivalists known to have participated in the revival
of 1233 or listed by Salimbene as preachers of the Ancient Way. Stories
that can be dated closer to the event described are favoured over those
more distant. All these stories were, of course, subject to stereotyping by
those retelling them, but I have not discounted a stereotypical story
merely because it is stereotypical. The regular repetition of a particular
topos can say much about what observers expected of a preacher. In short,
this study is as much concerned with popular perceptions of the preachers
as with their actual personalities or actions. In addition to the scattered
stories told of revivalists in the literature produced for internal con-
sumption by the mendicant orders, other sources have provided a rich
mine for the way in which the revivalists were viewed. Here I have
drawn on papal and imperial correspondence, contemporary legislation on
preaching by the Dominican order, and references found in lay writers.[59]

   No study of revivalism can avoid touching in some way on the sermons
that were preached. It is unfortunate that no sermons survive from
the known preachers of 1233. On the other hand, there does exist
one autograph manuscript containing sermons of Ambrose Sansedoni.
Other mendicant collections from thirteenth-century northern Italy have
provided some useful insights into possible themes in revivalistic
preaching.[60] In practice this material is far less useful for reconstructing

Peter of Verona, the life compiled by Ambrogio Taegio as the *Legenda Beatissimi Petri
Martyris ex Multis Legendis in Unum Compilata* (AS, 12: Apr. III; Paris, 1867), 694–727.
The last presents textual problems that will be treated later.

[59] The two contemporary lay observers of mendicant preaching in north Italy,
Boncompagno of Signi and Guido Bonatto, were very critical of its excesses, as are the
references to it in the legislation on preaching by the Dominican Order: *Acta Capitulorum
Generalium Ordinis Praedicatorum*, i, ed. Benedictus Maria Reichert (MOFPH 3; Rome,
1898); the *Acta Capitulorum Provincialium Provinciae Romanae (1243–1344)*, ed. Thomas
Kaeppeli, in 'Acta Capitulorum Provinciae Lombardiae (1254–1293) et Lombardiae
Inferioris (1309–1312)', *AFP* 11 (1941), 140–67; and the *Acta Capitulorum Provincialium
Provinciae Romanae (1243–1344)*, ed. Thomas Kaeppeli and Antoine Dondaine (MOFPH
20; Rome, 1941).

[60] I have drawn on the following mendicant sermons or preaching-aids traceable to north
Italy in the early to mid-1200s: Luke the Lector, OFM, *Sermones de Adventu et Festivis*,

the revival than one might think. The preacher was probably better seen than heard, and much of his impact came from his actions and tone rather than his specific words. As shall be seen in Chapters 4 and 5, the stories told of the revivalists and their activities indicate that their popularity flowed from perceptions about them that had little to do with their words from the pulpit. But the extant sermons that can be linked to mendicant preaching in early to mid-thirteenth-century Italy do confirm popular perceptions.

Unlike modern revivalism, which tends to emphasize the internal religious conversion of the participants, that of the Ancient Way emphasized an outer social dimension, the establishment of peace, the topic of Chapter 6. The peace-making and reconciliation was personal, communal, and inter-communal. The revival of 1233 could with justification be termed a peace movement.[61] The phrase is certainly accurate since forms of it are used by contemporaries. Salimbene, for example, in his first reference to the revival speaks of the Alleluia as 'a time of quiet and peace, as it was later to be called, when all the arms of war were far off, a time of gladness and delight, of joy and exultation, of praise and jubilation'.[62] These were enthusiastic, perhaps apocalyptic, words,

Padua, Biblioteca Antoniana MS 466; the anonymous *Sermones de Tempore et de Sanctis*, Padua, Biblioteca Antoniana MS 477; and the marvellous sermon-illustration collection by Bartholomew of Trent, OP, the *Liber Miraculorum Beate Marie Virginis*, in Bologna, Biblioteca Universitaria MS 1794, fos. 70ᵛ–108ʳ; On this codex and Bartholomew's other writings see G. Abate, 'Il "Liber Epilogorum" di fra' Bartolomeo da Trento O.P. in due codici rintracciati nella Biblioteca Antoniana di Padova', in *Miscellanea P. Paschini* (Rome, 1948), i, 269–92; I. Paltrinieri, 'Un nuovo codice di fra' Bartolomeo tridentino', *Aevum*, 20 (1946), 3–13; id. and G. Sangalli, 'Un opera finora sconosciuta: Il *Liber Miraculorum B.V.M.* di fra' Bartolomeo tridentino', *Salesianum*, 12 (1950), 372–97.

[61] As did Sutter and Vauchez.

[62] Salimbene, 70: '[Q]uoddam tempus, quod sic in posterum dictum fuit, scilicet tempus quietis et pacis, quoad arma bellica omnino remota, iocunditatis et letitie, gaudii et exultationis, laudis et iubilationis.' If Salimbene meant to recall his earlier Joachimism, he may have intended this phrase to reflect the eschatological eighth day of Church, the 'Mysterium Ecclesiae' of Joachim of Fiore, *Il Libro delle figure dell'Abate Gioachino da Fiore*, ed. Luigi Tondelli (2 vols.; Turin, 1953), ii, plate XIX. On Salimbene's Joachimism, see Delno C. West, jr., 'The Education of Salimbene of Parma: The Joachite Influence', in Ann Williams (ed.), *Prophecy and Millenarianism: Essays in Honor of Marjorie Reeves* (Harlow, 1980), 193–215. Seeing the Alleluia too much through Salimbene's eyes has led some scholars to call it 'Joachimite'. See e.g. Luigi Tondelli, 'L'anno del' "Alleluia" ', in *Il Libro delle figure*, i. 192–6. That Joachimite speculation triggered the Alleluia is highly unlikely since Joachim's works did not reach northern Italy until the 1240s. On this, see Delno C. West, jr., and Sandra Zimdars-Swartz, *Joachim of Fiore: A Study in Spiritual Perception and History* (Bloomington, Ind., 1983), 101. In spite of a diligent search, I find no evidence that the revival was triggered by the anniversary of Christ's passion in AD 33. Those seeking more information on Joachimism may consult Marjorie Reeves, *Joachim of Fiore and the Prophetic Future* (London, 1976); ead., *The Influence of Prophecy in the Late Middle Ages*

employed 50 years after the event, but it can be gathered from other chroniclers that they reflect the popular view of the revival.

The study of Alleluia peace-making leads directly to one of the Devotion's most striking features. For all its enthusiasm and emotion, thirteenth-century peace-making was pre-eminently a legal event, a contract or pact. The pact was drawn up through formal procedures of mediation and ratified by notarized instruments having the force of law. The study of peace-making in the 1200s must take these legal aspects of reconciliation into consideration. This work examines peace-making first as a social phenomenon and then as a legal one. To reconstruct the social aspects of peace-making, I have drawn first on narrative descriptions, found principally in the mendicant story-collections and hagiography. The significance of these stories has been drawn out by a reconstruction of the legal forms found first in the extant peace-instruments of 1233, and second in the procedures for peace-making laid out in early to mid-thirteenth-century notarial manuals and legal treatises.[63] These are, with the unfortunate exception of the great notarial manual of Rolandinus Passagerii, available in modern editions. To some extent also, I have drawn on municipal and corporate statutes and acts when these can throw light on peace-making. With the exception of the *Registro Grosso* and *Registro Nuovo* of Bologna,[64] these municipal acts are available in modern editions. Chapter 7 systematically analyses the legal forms used in peace-making during the mid-1200s.

Since peace-making was a legal, often statutory, procedure, it propelled the preacher into the role of legislator. In Parma, Modena, Bologna, and many cities of the Veneto, the friars took direct control of the communal governments and revised their statutes and laws at will. This activity frightened some thirteenth-century observers,[65] and has

(Oxford, 1969); and ead. with Beatrice Hirsch-Reich, *The* Figurae *of Joachim of Fiore* (Oxford, 1972).

[63] Useful for reconstruction of peace-making in the early 1200s are the earliest notarial manuals (1223–45). These include that of Rainerius Perusinus, *Ars Notariae*, ed. Ludwig Wahrmund (Quellen zur Geschichte des römisch-kanonistischen Processes im Mittelalter, 3; Innsbruck, 1917); that of Richardus Anglicus, *Summa de Ordine Iudiciario*, ed. Ludwig Wahrmund (Quellen zur Geschichte des römisch-kanonistischen Processes im Mittelalter, 2; Innsbruck, 1915); and the widely influential treatise on Roman-Canonical procedure by Tancred of Bologna, *Ordo Iudiciarius*, in *Libri de Iudiciorum Ordine*, ed. Friedrich Christian Bergmann, (1842; repr. Göttingen, 1965). I have also used the later manual of Rolandinus Passagerii (c. 1255), *Summa Totius Artis Notariae* (2 vols.; Venice, 1588).

[64] *Registro Grosso*, Bologna, Archivio di Stato MS; *Registro Nuovo*, Bologna, Archivio di Stato MS.

[65] See e.g. Maurisio, *Cronica Dominorum*, 33–4, who viewed John as a power-hungry politician.

attracted the attention of modern scholars.[66] If one wishes to understand the revivalists as legislators, their legislation must be carefully placed within the context of their principal political role, that of peace-maker. It must also be contrasted with ordinary communal statute-making and legislation. In Chapter 8, this study focuses on the one extant set of municipal statutes where the work of a revivalist of 1233 exists in mass, that of Parma.[67] One of the functions of the council in a north Italian city was to pass municipal statutes, and these were usually collected and preserved. They provide a firsthand look at the business of municipal government. For comparison with the reforms at Parma, I have also drawn on the contemporary statutes of Bergamo, Bologna, Monza, Padua, Treviso, Vercelli, Verona, and Vicenza.[68]

The second part of this book, then, consists of five chapters, each of which focuses on a particular distinctive aspect of the revival and its practitioners. The order reflects the progressive stages of the revival itself—audience-creation leading to miracle-working, both resulting in peace-making (considered first sociologically and then legally), the ratification of which is statutory reform. The Conclusion will suggest some reasons why the revival developed as it did, and why John, and with him the Great Devotion of 1233 itself, came to such rapid and utter ruin. It will also suggest the kind of legacy the revivalists of the Ancient Way bequeathed to their more famous successors such as Bernardino and Savonarola.

[66] In particular, Vauchez, 'Une campagne', which was the first attempt to analyse this legislation.

[67] Edited in *Statuta Communis Parmae Digesta Anno 1255* (Monumenta Historica ad Provincias Parmensem et Placentinam Pertinentia, 1; Parma, 1856).

[68] Bergamo: *Antiquae Collationes Statuti Veteris Civitatis Pergami*, ed. Giovanni Finarzi (Historiae Patriae Monumenta, 16: Leges Municipales, II; Turin, 1876); Bologna: *Statuti di Bologna dall'anno 1245 all'anno 1267*, ed. Lodovico Frati (3 vols.; Bologna, 1869–77); the *Statuti delle Società del Popolo di Bologna*, i: *Società delle Armi*, ed. Augusto Gaudenzi (Fonti per la storia d'Italia: Statuti, secolo XIII; Rome, 1889); fragments for Monza: in *Memorie storiche di Monza e sua corte*, ii: *Codice diplomatico*, ed. Anton-Francesco Frisi (Milan, 1794); Padua: *Statuti del comune di Padova dal secolo XII all'anno 1285*, ed. Andrea Gloria (Padua, 1872); Treviso: *Gli statuti del comune di Treviso*, ii: *Statuti degli anni 1231—1233—1260–63*, ed. Giuseppe Liberali (Monumenti storici pubblicati dalla Deputazione di Storia Patria per le Venezie, NS, 4; Venice, 1951); Vercelli: *Statuta Communis Vercellarum ab Anno MCCXLI*, ed. Giovambatista Adriani (Historiae Patriae Monumenta, 16: Leges Municipales, II; Turin, 1876); Verona: *Liber Juris Civilis Urbis Veronae*, ed. Bartolomeo Campagnola (Verona, 1728), and *Gli statuti veronesi del 1276 colle correzioni e le aggiunte fino al 1323*, i. ed. Gino Sandri (Monumenti storici pubblicati dalla Deputazione di Storia Patria per le Venezie, NS, 13; Venice, 1940); Vicenza: *Statuti del comune di Vicenza 1265*, ed. Fedele Lampertico (Monumenti storici publicati dalla Deputazione Veneta di Storia Patria, serie seconda: Statuti, I; Venice, 1886).

PART I
THE GREAT DEVOTION OF 1233

I
The Coming of the Preachers

DURING the fourth decade of the thirteenth century, northern Italy faced an unprecedented series of natural disasters. Bad weather and consequent poor harvests had begun in 1230, when a flood of the Po sent people scrambling to the tree-tops for safety.[1] Then, in August 1232, a plague of locusts and grasshoppers attacked the crops. The pests returned annually for 3 years, ravaging the region's vines until nothing remained but the roots.[2] The winter of 1234 was among the most bitter in recent memory. The cold devastated vines, olives, and fig trees. Those travelling the main road from Bologna to Padua could cross the frozen Po by horse or carriage when they reached Ferrara.[3] The cold was so fierce that year that the wolves abandoned the hills and entered the cities in search of food.[4] At the height of these disasters, in 1233, the result was, if not starvation, widespread want. Even the wealthy Bolognese felt the pinch as prices skyrocketed: they forwent wine and offered their wedding toasts with water.[5] There were many who must have looked to heaven for, if not relief, at least some explanation of these acts of God.

In the same year, 1233, as the warmth of spring finally arrived, the beleaguered citizens of Parma were afforded an unexpected and unusual diversion. There appeared in their midst, from where it is not clear (some said from Rome, others from the Valley of Spoleto), a bizarre lay preacher.[6] He called himself Brother Benedict, but the city knew him as 'Fra Cornetto', that is 'Brother Horn', since he punctuated his preaching with blasts on a trumpet. His demeanour and habit recalled an earlier age. He wore a tunic bound with a leather girdle, and over that he threw a long black cloak marked front and back with a red cross extending from

---

[1] Varignana, 98: '1230—Uno diluvio de acqua fuo grandissimo. Per lo quale gli uomini conveneno fugire a l'albori, se volseno scanpare la furia de l'acqua.' Especially hard hit were the cities of Ferrara, Padua, Mantua, and Cremona.

[2] Varignana, 101: '[R]odevano l'erbe dalla vetta infino alla radize.'

[3] CCB, 103–4.

[4] *Memoriale Potestatum Regiensium*, ed. Lodovico Antonio Muratori, (RIS 8, col. 1108).

[5] Leandro Alberti, *De Viris Illustribus Ordinis Praedicatorum* (Ferrara, 1516), fos. 183–4.

[6] *Ann. Parm. Maj.*, 668; Sutter, 29–31; and I. Walter, 'Benedetto', *Dizionario biografico degli Italiani* (1966).

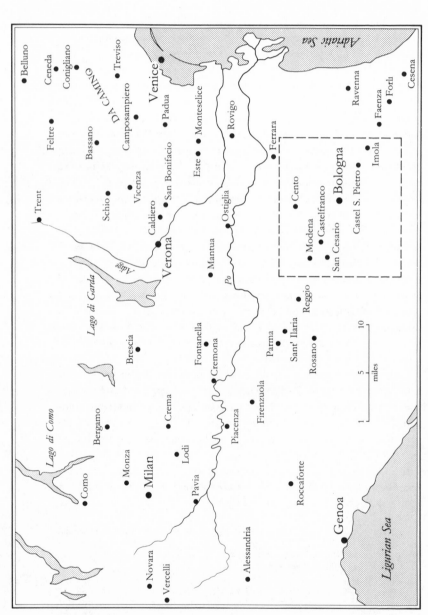

MAP 1. Northern Italy. For the *contado* of Bologna (boxed area), see Map 2.

his neck to his feet. This cloak resembled a priestly chasuble; on his head he wore an Armenian cap, and from his chin flowed a long black beard. He wandered from church to church and square to square, preaching, sounding his horn, and attracting a great following of children.

Brother Benedict might have disappeared and been forgotten, like so many other wandering medieval preachers and religious eccentrics. After all, he never founded a religious order and, as a layman, he lacked a community of learned clerics to celebrate his memory. He was known as a friend of the Franciscans, but the now increasingly comfortable followers of Saint Francis probably never saw him as anything more than a harmless enthusiast. Nevertheless, Benedict must have possessed a peculiar, arresting quality since he seems to have attracted the attention of two chroniclers. He seems to put in an appearance in the chronicle of Richard of San Germano, who apparently records his preaching in the Valley of Spoleto.[7] His other appearance, as luck would have it, was in the chronicle of that very chatty Franciscan, Salimbene de Adam of Parma.[8]

For Salimbene, Benedict was no ordinary lay preacher; he was a second John the Baptist sent 'to prepare a perfect people for the Lord'. This characterization seems accurate. He arrives, seemingly from nowhere, electrifies the people of Parma with his preaching, and then seems to disappear. There is no evidence that he co-operated with the later preachers of the Devotion of 1233.[9] Salimbene tells us that he himself often went to see the strange preacher haranguing the crowds from the uncompleted walls of Parma's episcopal palace. He listened to the preacher play 'both sweetly and horribly' on his trumpet and watched him lead throngs of children through the streets. Benedict's followers carried

---

[7] Richard of San Germano, *Chronica*, ed. Carlo Alberto Garufi (RIS² 7:2; Bologna, 1935–7), 370: 'Eodem mense frater B., vili confectus tegmine, tamquam de ordine fratrum Minorum ad S. Germanum veniens cum cornu quodam convocabat populum, et alta voce cantabat tertio Alleluia, et omnes respondebat: Alleluia.' This identification of him as a Franciscan may be an understandable error, but Eduard Winkelmann, *Kaiser Friedrich II*, ii (Leipzig, 1897), 439–42, reading 'frater I'., denies that this reference refers to Benedict at all.

[8] Salimbene, 70–2.

[9] Brother Benedict's lack of direct association with the mendicant preachers who dominated the Devotion of 1233 may be the element of truth in the suggestion of Fumagalli, 257–72, that there were two revivals during 1233. His perspective is vigorously argued by Daniela Gatti, 'Religiosità populare e movimento di pace nel'Emilia del secolo XIII', in *Itinerari storici: Il medioevo in Emilia* (Carpi, 1980), 80, 92–3, 95. For reasons to be explained in Ch. 6, the two events cannot be distinguished, following Fumagalli, as a 'peace' revival and a 'penance' revival, but it does seem likely that Benedict's revival was essentially a lay movement, while what followed him was led and promoted by clerics, that is, Dominican and Franciscan preachers.

lighted candles and tree-branches and shouted or sang the praises of God and the Virgin in the vernacular. Even after 50 years, Salimbene could remember the words of their songs, which included a troped version of the Hail Mary and praises of the Trinity interspersed with Alleluias.[10] Since Salimbene was himself a mere youth of 12 years in 1233, he probably not only observed Benedict's processions, but also participated in them.

The restless Benedict did not stay in Parma; by June of 1233, when the Devotion of 1233 was in full swing, he appears to have returned south to San Germano, if Richard's supposed reference to him and his preaching is correct.[11] His journey continued south. When he finally disappears from sight in 1236, his campaign has reached Apulia. A report of events there eventually reached the Emperor Frederick II, who forbade Benedict to preach and launched a letter of protest to the pope.[12]

Benedict's enthusiasm enkindled no great upsurge of devotion in the south, but in Emilia he was no longer needed to continue the revival he had sparked. The young Salimbene may have found this horn-blowing preacher entertaining, but it was not for that reason that he portrayed him in the chronicle. For Salimbene, Benedict was a *new* John the Baptist, making straight the path of the Lord. He was the forerunner of events that would prove far more significant in the religious and political life of Parma and northern Italy.[13] Perhaps it was this peculiar lay preacher's unexpected success that inspired the host of Dominican and Franciscan imitators who appeared in north Italy during the rest of 1233.

---

[10] Salimbene, 72, gives us this version of their song: 'Ave Maria clemens et pia, | Gratia plena, virgo serena! | Dominus tecum, et tu mane mecum! | Benedicta tu in mulieribus | que peperisti pacem hominibus, | et angelis gloriam! | Et benedictus fructus ventris tui | qui, coheredes ut essemus sui, | nos fecit per gratiam.' This would seem to be a translation into Latin of one of the earliest vernacular *lodi*. The *Hail Mary* seems to have been sung antiphonally. See, Fernando Liuzzi, *La lauda e i primordi della melodia italiana* (2 vols.; Rome, 1934), i.

[11] Richard of San Germano, 31; Salimbene, 71, also speaks of him as active at Pisa. Chronicles for Pisa make no reference to him, however; see I. Walter, 'Benedetto'.

[12] Letter of Frederick II to Gregory IX, 20 Sept. 1236, H.-B., ii, 905: 'Sic in Apulia quidam frater assumens a puerorum simplicitate primordium iam multos ibidem sub vexillo proprio congregaret. Hoc etsi boni forte sit zelus, quia tamen mali speciem non evitat, procul dubio iussimus inhiberi predicari.' For the general background of Frederick II's role in northern Italy at the time of the Alleluia, see Ernst Kantorowicz, *Frederick the Second, 1194–1250*, trans. E. O. Lorimer (New York, 1957), 388–99. But be warned, Kantorowicz seems to depend almost completely on Sutter and considers John of Vicenza, the principal preacher of 1233, to be staunchly anti-imperial—a very dubious assumption.

[13] Salimbene, 71: 'Quasi alter Iohannes Baptista videbatur, qui precederet ante Dominum et pararet Domino plebem perfectam.'

Beginning in the central cities of the Emilian Way, preachers imitating the methods of Benedict fanned out northward into the Veneto and the March of Treviso, and westward into Lombardy.[14] They generated a tempest of religious enthusiasm unmatched in Italy until the arrival of the flagellants over 25 years later.[15]

The revivalists were, to a man, from the mendicant orders. Some were Franciscans, members of the order so dear to Brother Benedict; others were Dominicans. They preached morning, noon, and night, and organized devotions for the faithful. Crowds came from the countryside; municipal corporations assembled under their banners; all marched through the cities carrying lighted candles and tree-branches. As they marched they sang songs and hymns, 'more with the voice of God, than of men'.[16] Almost immediately the revival received a name: 'The Alleluia'. The chroniclers took it from the crowd's practice of singing that chant as they came to hear the preachers. Other names followed: 'The Great Devotion of 1233' and the 'Great Devotion of the Friars Preachers.' These names all refer to the same outpouring of enthusiasm.[17]

The two most important leaders of the revival after the departure of Benedict were the Franciscan Gerard of Modena and the Dominican John of Vicenza. In the early stages of the revival one can discern a division of labour between the Franciscans and Dominicans along the lines of inter-communal alliances. During the inter-communal wars of the 1220s and 1230s, Parma had been an ally of Modena, the home town of Gerard who was active in both; while Reggio, the site of the preaching of a Dominican, Jacopino, lay trapped between the hostile cities of Parma and Modena. Reggio was allied with Bologna, which became the base of operations of another Dominican, John of Vicenza. As the revival continued into the summer of 1233, the preachers, perhaps under the guidance of John, began to co-ordinate their efforts, and this division of labour disappeared.

[14] Fumagalli, 257–72, contrasts the joyful peace-preaching of Benedict and the Franciscans with the penitential preaching of the Dominicans. On the unity between peace-preaching and penance-preaching in this period, see Marie-Humbert Vicaire, *Dominic et ses prêcheurs*, 2nd edn. (Paris, 1977), 284–5.

[15] A comparison made by Salimbene, 467, himself. The revival received attention throughout Europe: in France, positively, from Alberic of Trois Fontaines, *Chronica*, ed. P. Scheffer-Boichorst (MGH.Ss. 23), 933, and in England, very negatively, from the Benedictine Matthew Paris, *Chronica Maiora*, ed. Henry Richard Luard (Rolls Series 57, 7 vols.; London, 1872–83), iii. 128.

[16] Salimbene, 70.

[17] As I have noted, Salimbene uses all three names interchangeably.

THE REVIVAL SPREADS WEST

Parma can claim the honour of giving birth to the revival. Benedict's successor as leader there was friar Gerard, born Maletta. He belonged to the wealthy and powerful Boccabadati clan of Modena and was an early convert to the Friars Minor. He had responded to the preaching of Saint Francis in Lombardy at some time in the 1210s. He was a dynamic, restless sort, little satisfied with remaining in one place.[18] Strong-willed and independent, he inherited from his family a predilection for the pro-imperial party. This predilection was something that later, on at least one occasion, enraged Pope Innocent IV's brother-in-law, Bernard dei Rossi, who believed that Gerard was prejudiced against the claims of Bernard's family.[19]

Gerard had first preached in Modena, his home town, perhaps while Benedict was still active in Parma. There he joined, or perhaps organized, unnamed Dominicans and Franciscans for preaching and processions.[20] One year earlier Modena had seen its *podestà*, Gabriel dei Conti of Cremona, murdered and the assassins brutally punished. The city immediately divided into factions, who fought in the streets. In early 1233, the commune begged Gerard to mediate.[21] He did this so effectively that all but five of those exiled during the unrest returned. The thankful population expressed its appreciation to God and the friars by helping construct the new Franciscan Church of Sant'Agnese.

From Modena Gerard went to Parma, taking up where Benedict had left off. Other preachers had prepared the way for him. There, by early January of 1233, suspicion of malconduct by the bishop had already caused Pope Gregory to commission a Dominican, Guala, the bishop of Bergamo, to investigate such accusations with the assistance of the abbot of Cerreto.[22] It had been reported to the pope that the bishop, 'acting more like a wolf than a shepherd,' had teamed up with officials of the

---

[18] Salimbene, 75: 'Totum mundum circuire volebat.'

[19] Ibid.

[20] *Ann. Vet. Mut.*, 60: 'Praedicatores fratrum Minorum et S. Dominici crescere ceperunt cum confalonibus et crucibus . . . de anno 1233 multae predicationes factae fuerunt.' On him, see Gustavo Cantini, 'L'apostolato dei beati Gherardo Boccabadati e Leone Valvassori da Perego francescani e la devozione dell'Alleluia', *SF*, 3rd ser., 9 (1938), 335–53; Sutter, 33–9, whose brief analysis of Gerard's statute reforms has been replaced by Vauchez, 519–49. I shall re-evaluate Gerard's reforms in Ch. 8.

[21] Salimbene, 75; *Ann. Parm. Maj.*, 668.

[22] Letter of Gregory, 12 Jan. 1233, Auvray, i. 603–4, letter 1036 (Potthast 9071); text in the 18th-cent. historian of Parma, Ireneo Affò, *Storia di Parma* (4 vols.; Parma, 1793), iii. 362–4.

commune to subjugate both clergy and laity and extort money from them. And to what purpose? Nothing less than constructing himself the splendid new palace from the top of whose unfinished walls Benedict had given his sermons! As Guala investigated this matter and heard the citizens' complaints, he no doubt joined in the preaching that was then convulsing the city. It is to be understood that other Dominicans came along with Guala, and that they too joined their voices to that of Benedict, who was still at work in the city.[23] In any case, when Gerard arrived in the early spring, he found the city already in a state of religious exaltation. He soon proved himself a skilled preacher. He claimed for himself the grace of divine guidance, which occasionally manifested itself in visions, received while preaching. It was not only God who appeared to him. Sometimes he saw John of Vicenza, sometimes the preacher Jacopino of Reggio, and at other times someone else. Gerard, like so many other revivalists of 1233, became notorious for his visions and miracles.

Gerard's stay in Parma was short, perhaps 6 or 7 months. During the spring he established his reputation. By early July, he had not only galvanized the populace by his preaching but convinced them to appoint him to negotiate between the city's feuding families and factions.[24] Commissioned as arbitrator for about 2 months (from 15 July to 29 September), he issued decree after decree, not only putting to rest grudges and bad feelings, but deciding lawsuits and ruling on points of law. He received authority to absolve anyone who had broken the peace before 28 July 1233. Not satisfied with the office of mediator, he requested complete control of the municipal government and the authority to rewrite city statutes. They named him *podestà*, with extraordinary powers.[25] He took the municipal statutes in hand, amending, abolishing, and adding whatever seemed good in his sight. The 1255 codification of Parma's statutes preserved most of this legislation for us to read today. When, in September, Gerard relinquished his powers and moved on, the populace expressed their appreciation to the mendicant preachers by filling in a large ditch that had disfigured the property adjacent to the

---

[23] Affò, iii. 153, reports preaching by the Dominicans Jacopino of Reggio, Bartholomew of Breganza, and John of Vicenza. If he is correct, this confirms the centrality of Parma during the early spread of the Devotion.

[24] His commission is found in Parma Stat., 312.

[25] Salimbene, 75: 'Huic tempore illius devotionis predicte Parmenses totaliter dominium Parme dederunt, ut eorum esset potestas, et concordaret eos qui guerras habebant ad pacem.'

Dominican church.[26] By that time the revival that Gerard promoted had already spread far afield.

North of Parma, at Piacenza, the revival began with the arrival of a flamboyant Franciscan, Leo de' Valvassori of Perego.[27] Unlike the humble Benedict, he came from the highest Milanese aristocracy. Leo was above all a showman. Rumours circulated that his birth had been heralded by portents and that he had been privy to visions.[28] Cleric though he was, he showed himself fearless and feckless in battle. Later in life, after becoming archbishop of Milan, he once bolted alone from the security of the city to face down the attacking troops of the Emperor Frederick. Standing on the field, banner in hand, he punctuated his barrage of insults against the imperial forces with shouts of ridicule at the Milanese holed up behind the city walls.[29] Leo made his reputation during the Devotion of 1233. After that year, preferments came quickly: he became minister provincial of the new Franciscan Province of Lombardy by the end of that year; then, in 1241, he received the archbishopric of Milan.[30]

In 1233, Piacenza had need of a peace-preacher of Leo's stripe. The struggle between the traditional leaders of the communal government and the popular party, a conflict typical of the early thirteenth-century cities of north Italy, had come later to Piacenza than to most other cities of Emilia. In the spring of 1232 the captain of the people, William De Andito, refused to obey the *podestà*, Guifredo de' Pirovali.[31] Armed populars burst into a meeting of the city council and, in the revolt that followed, drove Guifredo from the city. The old leadership of the commune capitulated. A government led by a committee of four *podestà* provided a compromise solution. In this division of power the popular party received control of half of the commune's property and offices. The rise of the popular party led almost immediately to war.[32] On 9 June of the same year, in alliance with the Cremonese, the popular militia

---

[26] *Ann. Parm. Maj.*, 668.

[27] On whom see: Cantini, 335–53; Williell R. Thompson, *The Friars in the Cathedral: The First Franciscan Bishops, 1226–1261* (Toronto, 1975), 93–4; Sutter, 25–9.

[28] Fiamma, *MF*, 678.

[29] Salimbene, 35.

[30] P. Sevesi, 'Documenta hucusque inedita saeculi XIII pro historia almae fr. Minor. Provinciae Mediolanensis (seu Lombardiae)', *Archivum Franciscanum Historicum*, 4 (1911), 656–7, 259–61. He seems to have been a remarkably unsuccessful bishop, probably because of his unruly temper, W. R. Thompson, *Friars*, 94. He died in exile at Legnano, Fiamma, *MF*, 678–9.

[31] *Ann. Plac. Guelf.*, 454, supplemented from Mussi, 461; and, more briefly, in *Chronicon Placentinum*, ed. Lodovico Antonio Muratori (RIS 16, col. 459).

[32] *Ann. Plac. Guelf.*, 454.

invaded Fiorenzuola and Roccaforte. The troops behaved savagely, devastating fields and vineyards, and finally running amok. They headed south-east into the hill regions dominated by Parma. Finally, they burned a monastery at Rosano. Moving northward towards Cremona, they devastated the entire region from Fiorenzuola to Fontanella and reportedly left it depopulated.

Tensions between the two factions were running high when Leo arrived, probably in April of the following year. His arrival triggered a rash of marvellous reports, including one of a miracle in the Dominican church.[33] Leo received a commission to arbitrate the dispute between the two factions. He exercised his mandate at a general assembly in the piazza of the Duomo, during which twenty members of each faction exchanged the kiss of peace. In July, after negotiations lasting about 3 months, Leo announced the terms of the reconciliation and named a new *podestà* who was acceptable to both factions, Lantelmo Mainerio. He arranged an equal division of government powers between the populars and the old governing class. The populars then assembled in the piazza of the Duomo and formally dissolved their militia. On the same day the old leading families received permission to refound their corporations.[34] It seems that a new balance of power had been achieved.

The reconciliation, so gloriously initiated, did not endure. By August, the city standard-bearer was leading a popular revolt against the traditional government. The populars prevailed during the ensuing civil war, and exiled most supporters of the old regime from the commune. Piacenza was no longer a fertile ground for revivalism or peace-making: when the learned Dominican Roland of Cremona arrived in the autumn, preaching against heresy, a mob stoned him and drove his retinue from the city. During the mêlée, one of his followers, a monk of San Savino, was murdered.[35] Life at Piacenza was back to normal.

---

[33] Mussi, 461: 'Eodem anno apparuit in ecclesia fratrum Predicatorum miraculum de Bussolla.' Since this is a unique reference, the nature of this miracle is unknown. The meaning was already forgotten by the 18th cent., when Cristoforo Poggiali, *Memorie storiche di Piacenza* (Piacenza, 1758), v, 172, wrote of the word, 'le quali parole un'enimma contengono per me insolubile'. A 'bussolla' in Italian could be, among other things, a vestibule door or an alms' box.

[34] Mussi, 461.

[35] Ibid.; On 15 Oct. 1233 Pope Gregory IX wrote to the bishop of Piacenza seeking information about the riot, Rodenburg, i. 556 (Auvray, i. 858: letter 1560). On 22 Oct. he wrote to the archdeacon of Novara, commissioning him to investigate the matter and name a *podestà* favourable to the Church, Rodenburg, i. 559 (Auvray, i. 862: letter 1569). The matter seems to have dragged on into the following year: see letters of 10 Dec. 1233, Auvray, i. 883–4: letters 1606, 1607, and of 15 Feb. 1234, Auvray, i. 983–4, letters 1795 (Potthast 9404), 1796.

This fiasco late in the summer should not mislead us into thinking that
the revival had lost its appeal so quickly. From Piacenza it continued to
spread west into Lombardy. Nevertheless, Milan, the region's most important city, remained untouched—unless some unknown preacher of the
movement sparked the outbreak of persecution directed against heretics
there during 1233.[36] During the summer, Franciscans revised statutes in
the cities of Vercelli and Monza;[37] miracle-working and peace-preaching
were reported in Como. In this last city, an unnamed Franciscan preacher
from Padua correctly predicted divine retribution against workmen
whose hammering had interrupted his sermon.[38] Since the Dominican
preachers Guala of Brescia and Bartholomew of Breganza[39] were natives of
Lombardy, one can surmise that they carried the revival north and east
from Emilia. Unfortunately, no chronicler bothered to record any par-
ticulars of their activities.

[36] *Memoriae Mediolanenses*, ed. Philipp Jaffé (MGH.Ss. 18, 402): 'Mediolanenses
inceperunt comburere ereticos.' This is repeated in Fiamma, *MF*, 672, which names the
podestà, Oldrado Grosso, as the leader of the persecution. This persecution is often linked
with the Dominican Peter of Verona. Bernardino Corio, *Mediolanensis Patria Historia* (Milan,
1503), i. fo. 7ᵛ, is the earliest to make the connection, giving Peter the title of Papal
Inquisitor. The only evidence he gives is a letter of 22 May 1231 by Gregory IX to the
bishop and commune of Milan—never mentioning Peter's name—concerning the
suppression of heresy, Auvray, i. 419 (letter 659), and three documents concerning Peter
(doc. 1, 2, and 3). Corio's dating of these documents is certainly in error. Antoine
Dondaine, 'Saint-Pierre-Martyr: Études', *AFP* 23 (1953), 71–3, taking into consideration
the corrections of Corio's dating by G.-G. Meersseman, 'Les Confréries de Saint-Pierre-
Martyr', *AFP* 21 (1951), 113–16 excludes the possibility of Peter's having received the
office of inquisitor before 1251.

Sutter, 24–5, picked up the old error from Henry C. Lea, *History of the Inquisition in the
Middle Ages* (3 vols.; London, 1886), ii. 207. In support of it he can cite only *Ann. Bergom.*,
810: 'Anno 1233. claruit Ioannes de Bononia et frater Petrus de Mediolano.' This certainly
refers to the young friar Peter's first assignment to the Dominican house in Milan; see
Borselli, *CMG.*, fo. 21ʳ (edited in App.). The persecution began only late in 1233, after 10
Dec., when Pope Gregory IX wrote to encourage the Milanese in the persecution of heretics,
BOP, i. 105. Later he gave papal protection to the defenders of the faith, Rodenberg, i.
459.

[37] I shall examine this legislation and its possible relation to the revival in Ch. 8.

[38] Salimbene, 74.

[39] Salimbene, 72, mentions Bartholomew vaguely, and later, ibid. 595, lists the leaders
of the revival: 'Isti fuerunt frater Iohannes de Vincentia, qui Bononie miracula faciebat et de
ordine Predicatorum erat. Item frater Bartholomeus similiter de Vincentia de ordine
supradicto, qui Parme miracula faciebat. Frater Iacobinus de Parma, qui in Regio miracula
faciebat et ideo de Regio dicebatur, et de ordine fratrum Predicatorum erat. Item frater
Ghirardus de Mutina ex ordine fratrum Minorum, qui totam Italiam circuibat predicando et
miracula faciendo. Frater Leo archiepiscopus Mediolanensis ex ordine Minorum similiter,
qui in Mediolano optime predicabat.' Except for this passage, the descriptions of Brother
Benedict, the papal letter to Guala, and the late report of Affò, no other documents exist
listing the names of Alleluia preachers.

THE REVIVAL SPREADS EAST

John of Vicenza was the one leader of the revival who so impressed his contemporaries that they recorded enough about him to allow the reconstruction of something resembling a biography.[40] It fell to him to carry the Devotion out of Emilia to the east and north.

Although his role in the early days of the revival was secondary to those of Benedict and Gerard, John eventually became the leader of the Great Devotion throughout northern Italy. Since there is enough evidence for a full narrative of this activity, and it gives us the best (and only) portrait of an Alleluia revivalist in action, the next two chapters are devoted to him alone. His connections with the other preachers of the revival in its early stages and his role in the early spread of the movement are so important, however, that he must be introduced here.

Before the Alleluia, John of Vicenza is, for us, a shadowy figure. Information about this part of his life is scarce and reconstruction of his youth, background, and education hypothetical.[41] Nevertheless, what is known is highly suggestive of the way in which youthful experiences and religious formation could affect the attitudes and outlook of a medieval revivalist.

John was born into a moderately well-to-do family of Vicenza about the year 1200.[42] He was the son of one Manelino, a lawyer (*causidicus*) and one-time treasurer of the city. His family was undistinguished in lineage.[43] They would have belonged to the popular faction. John's

---

[40] For the scholarship, mostly hagiographic and of little value, on John of Vicenza before 1772, see Paolo Calvi, 'Angelogabriello di Santa Maria', in *Biblioteca e storia di scrittori di Vicenza*, i (Vicenza, 1772); for 1772 to 1891, see Sutter, 177–84. Basilio da Schio, *Biografia del beato Giovanni da Schio* (4 vols.) Vicenza, Biblioteca Comunale MSS E. J. 12–15 collected, without system, nearly every reference to John. Carl Sutter's *Johann von Vicenza und die italienische Friedensbewegung im Jahre 1233* (Freiburg, 1891), translated, abridged, and edited by Maria, Gelda, and Olga da Schio as *Giovanni da Vicenza e l'Alleluja del 1233* (Vicenza, 1900), is the most recent biography of John. He supplements the bibliography in Calvi. Daniel A. Brown, 'The Alleluia: A Thirteenth Century Peace Movement', *Archivum Franciscanum Historicum*, 81 (1988), 9–14, has provided a narrative in English of John's activities that follows Sutter, setting the revival in the wider context of peace-making.

[41] On John's origins, see Sutter, 42–62.

[42] In a document issued by John during the Alleluia, found in Savioli, iii. 128, he wrote, 'Ego, fr. J. de Bononia nunc qui olim fui de Vicentia oriundus.' This is confirmed by Maurisio, 31; *Ann. Brix.*, 818; Salimbene, 38; and *Ann. Vet. Veron.*, 91. The date 1200 for his birth was first proposed by Francesco Barbarano, *Historia ecclesiastica della città, territorio e diocesi di Vicenza* (Vicenza, 1649), 51, and is accepted by Sutter, 47.

[43] Sutter, 44–7, refuted the tradition that John belonged to the family of the counts of Schio. Maurisio, 31, says he was 'nacione plebeius'. John's noble origin first appeared in the 17th cent. as a result of a series of arbitrary deductions from a series of individuals with the

earliest memories were of the devastating civil war that rent Vicenza during the first decades of the thirteenth century.[44] As in so many other cities of north Italy, the struggle was over control of the communal government. By the late twelfth century the citizens of the commune had already divided into two factions. One was led by the Vivaresi clan, the other by the family of the counts of Vicenza. The leadership of both parties was drawn almost entirely from the nobility. Each faction repeatedly attempted to assure the appointment of a single *podestà* favourable to it, but, for the most part, they had to be satisfied with choosing a *podestà* with authority over their own party alone. The city chafed under a succession of two-headed governments.

In the early 1200s the popular elements in the city united and seized control of the commune through the appointment of the Milanese William Pusterla as *podestà*. There were now three factions in the city. When William demanded hostages and security from the leading nobles, the Vivaresi faction revolted and expelled him. Paradoxically, during the political struggles that followed the expulsion, the popular faction generally supported the Vivaresi, who were seen as the best bulwark against the powerful da Romano family. This family had its power-base in the countryside east of the city and had collected a large following of landless knights. The da Romano replaced the counts as the leaders of those opposed to the Vivaresi. By 1217, about the time of John's entry into the Preachers, Vicenza broke out in revolution. The populars successfully demanded one third of the city offices and then, in concert with the Vivaresi, and under the leadership of the noble *podestà* Uguccio dei

name 'da Schio'. At the beginning of the 14th cent., Godi, 10, describes John as 'Ioannes de Scledo'. Godi, 85, speaks of one Matteo da Schio as *podestà* of Lonigo, and Giambattista Pagliarini, *Cronaca di Vicenza*, ed. and trans. G. Alcaini (Vicenza, 1663), 271, speaks of a Vicenzo da Schio who was a follower of Ezzelino da Romano at Padua. In the early 17th cent. Georgio Piloni, *Historia* (Venice, 1607), fo. 104ᵛ, mentions a Martino da Schio as *podestà* of Belluno during the 1230s. Suddenly, in the 17th cent., Vicenzan writers gratuitously identify this Martino with John's father Manelino: Barbarano, *Historia ecclesiastica*, 51; Calvi, 'Angelogabriello', 29. It was left to the genealogist of the da Schio family, Basilio, to weave these references into a family history. A manuscript of Maurisio copied in 1455 shows us that at that date John's nobility was still unknown, see Maurisio, 31, n. 2. Here it is only a later hand that has erased the testimony that John was 'nacione plebeius'.

[44] Maurisio is the principal narrative source for this period of Vicenzan history. See also J. K. Hyde, *Society and Politics in Medieval Italy: The Evolution of the Civic Life, 1000–1350* (New Studies in Medieval History, London, 1973), 109–10; G. de Vergottini, 'Il "popolo" di Vicenza nella Cronaca ezzeliniana di Gerardo Maurisio', *Studi Senesi*, 48 (1939), 354–74; and Girolamo Arnaldi, *Studi sui cronisti della Marca trevigiana nell'età di Ezzelino da Romano* (Rome, 1963), 27–66.

Rambertini, they expelled the da Romano clan and any populars who had supported them. This alliance remained in power and, after 1229, allied with the anti-da Romano faction of nearby Verona. The da Romano were content to collect a following of landless knights and terrorize the countryside.

John's family would have belonged to the populars of Vicenza; as such, they would have allied with the Vivaresi, since that was the traditional popular affiliation. Since class and party affiliation seem to have been only loosely connected, and the populars were never strong enough to control the city to the exclusion of the older ruling class, John could draw a political lesson from the turmoil in his home town: government by a single faction was futile. The best solution was rule by an outsider who could preserve a system of government that would keep the leaders of all the factions in check. Pusterla had tried to do this but failed. Perhaps, in the hands of the right man, such a scheme would succeed.

Dominican traditions tell us that the young John went to study at Padua and there entered the Dominican order after hearing the preaching of Dominic.[45] Studies at Padua were a reasonable option for the son of a lawyer, and Dominic's preaching there in 1217 had sparked the foundation of a Dominican convent. The construction of the Dominican church had to wait until 1226, when it began under Prior Guido. It was complete in 1229, 4 years before the revival. Dominic may have preached again in the area before the Dominican chapter at Bologna in 1220.[46] So, if it was Dominic's activity that triggered John's religious conversion, his entry into the order would have occurred in either 1217 or 1220. A late report that John moved almost immediately to the Dominican house in Bologna is not unreasonable.[47] A son of the new convent in Padua would have passed to the central house of his province to begin his theological studies. If Salimbene's report that John was 'of little learning' is trustworthy, one might surmise that he rushed through his education to get down to his life's vocation, preaching.

[45] Valerio Moscheta, *B. Joannis Vicentii O.P., Professi Cenobii Paduani S. Augustini, Doctrina, Sanctitate, et Miraculis Insignis Praeclara Gesta* (Padua, 1590), ed. Joannes Baptista Sollerius (AS 28: Jul. I, 413; Paris, 1867); Alberti, *De Viris Illustribus*, 483; Giovanni Michele Piò, *Vite degli huomini illustri di s. Domenico* (Bologna, 1620), 53. Sutter, 47, correctly rejects the report of Barbarano, *Historia ecclesiastica*, 55, that John first studied at the (as yet non-existent) University of Vicenza.

[46] See Vicaire, *Dominic*, 301–2, 320–1.

[47] Moscheta, *B. Joannis Vicentii*, 415; followed by Sutter, 51; although this seems to be contradicted by Villola, 101, who tells us that in 1233 'Iohannes de ordine fratrum Predicatorum venit primo civitatem Bononie.'

By the late 1220s John was a proven religious. In 1231, he was elected prior of his mother convent, Sant'Agostino of Padua.[48] There, and in that capacity, after the death of Anthony of Padua, John received a commission from Gregory IX to serve on the panel to investigate the great Franciscan preacher's claim to sainthood. So, by the early 1230s, John was an established figure in the north Italian province of the Dominican order and in the church affairs of the region.[49] In Padua, John would have had the opportunity to observe something he had never seen before: a city at peace and generally free of factional strife. Padua had refused to ally with any of the magnate-led factions of the eastern Veneto. Its leaders remained aloof from both those who followed the da Romano and those who followed the anti-da Romano party of the duke of Este. This sagacious neutrality was the work of one man, Jordan Forzaté of the Benedictine order—the city's spiritual father and a reputed saint.[50] Although he never held any office other than that of prior of San Benedetto, the city fathers reportedly obeyed him in everything. This model of government would not have been lost on the young Dominican.

Perhaps he had already started mixing miracle-working and preaching, the combination that would later make him famous.[51] Certainly he was already a popular preacher. He had good models: St Dominic, Dominic's charismatic successor Jordan of Saxony, and the legendary Anthony of Padua.[52] The effects of his early preaching are unknown—no descriptions of it survive. Less than 2 years pass before the first reports. These come from Bologna in 1233, where it had fallen to John to carry the Great Devotion. We have already seen that Gerard began his revivalism at Modena, even before he came to Parma. John of Vicenza was responsible for Bologna and carried the revival to Faenza and the cities to the east.

By April, then, the Devotion had touched all the major cities along the Emilian Way from Piacenza to Faenza, with one exception, Reggio. There, the revival was slow in coming, perhaps because the early part of

---

[48] Moscheta, *B. Joannis Vicentii O. P.*, 472; Barbarano, *Historia ecclesiastica*, 61. Sutter, 50, rejects the late and often-repeated story that John served as lector in theology at Sant'Agostino in 1230. It was highly unlikely that theology was taught there in the 1230s: see Heinrich Denifle, *Die Universitäten des Mittelalters bis 1440* (Berlin, 1885), i, 280.

[49] We know nothing else of his activities in the late 1220s and early 1230s; see Sutter, 55–61.

[50] Vita in AS 32: Aug. II, 200–14; most recent biography: Italo Rosa, *Il beato Giordano Forzaté, abbate e priore di San Benedetto in Padova, 1158–1248* (Bresseo, 1932).

[51] That is, if we trust the dating of one miracle-story to 1231 by Th. Cant., 424.

[52] On Jordan's preaching, see Marguerite Aron, *St Dominic's Successor* (London, 1955), ch. 5; on Anthony, *Assidua*, 326–34.

1233 had seen the city in a state of civil unrest because of the outbreak of a feud. The feud arose from an otherwise trivial incident. Two children from important families of the city, Sigismund Malaguzzi and Hippolytus Ruggieri, were playing at war in the Piazza Comunale and were duelling with sticks. During the mock battle, Sigismund managed to poke out his playfellow's eye. Hippolytus ran off screaming and Sigismund escaped to the safety of his family tower. The two clans immediately converged on the square, words were exchanged, swords drawn, and a riot broke out. In no time the city had divided into two factions and men fought in the streets. Finally and with much difficulty, the *podestà*, Nicholas of Doara, brought about a reconciliation.[53]

The Devotion arrived under Nicholas' successor, Giliolo de Gente, Salimbene tells us, in the person of a Friar Preacher, Jacopino of Reggio.[54] Jacopino's arrival at Reggio must have followed his preaching elsewhere. He did not take up preaching in Reggio until after hearing John of Vicenza preach to troops of Bologna and Modena assembled on the field of battle. This occurred in late May or June.[55] By then Reggio must finally have been ready to hear words of peace.

Unlike the uneducated Benedict, the Dominican Jacopino had received a first-rate theological and rhetorical education.[56] To his training he added an instinct for organization and an appreciation of the impact co-ordinated preaching could have on its public. He and fellow revivalists soon began to direct the revival from regular planning and strategy sessions. To the devotions practised by Benedict, Jacopino added others that spectators later identified as typical of the revival. His processions left Reggio for the countryside. There they rendezvoused with others from the nearby village of Sant'Ilaria and the further-off city of Parma to hear his preaching of reconciliation and peace.[57] As Jacopino's activity

[53] *Doppelchronik von Reggio und die Quellen Salimbenes*, ed. Alfred Wilhelm Dove (Leipzig, 1873), 164; *Memoriale Potestatum Regiensium*, col. 1107; see also, Giancarlo Silingardi and Alberto Barbieri, *Storia di Reggio Emilia* (Modena, 1970), 33.

[54] *Doppelchronik*, 164; *Memoriale Potestatum Regiensium*, col. 1107—this laconic text seems to link Jacopino's preaching with that of John of Vicenza on the border between Bologna (Reggio's ally) and Modena (its enemy): 'inter castrum Leonem & Castrum Francum'; Salimbene, 35. See Sutter, 39–40.

[55] *Memoriale Potestatum Regiensium*, col. 1108: The rubric gives the date of this as 1233.

[56] Salimbene, 73: '[L]itteratus homo fuit et lector in theologia: fecundus, copiosus et gratiosus in predicationibus; homo alacer, benignus, caritativus, familiaris, curialis, liberalis et largus.'

[57] Jacopino continued to play a role in the politics of Parma after 1233: Parma Stat., 216, lists him along with the Dominican Peter of Verona and four Franciscans as witnesses to a government act.

intensified, rumours began to circulate that he was working miracles and receiving visions. What had begun as a local outpouring of piety, God had transformed into a miraculous intervention of Grace.

After Jacopino preached in the countryside, the crowds returned to the city and confirmed their sincerity by dragging stones, sand, and cement to construct swiftly a new Dominican priory. Construction formally began on 25 July 1233, the feast of St James, when Nicholas, the bishop of the city, consecrated the corner-stone. One might say that he and the people of the city had belatedly added the last link in the chain of revivalism across northern Italy. John of Vicenza had meanwhile already captured Bologna and was now on his way northward, carrying the revival to the Veneto and the March of Treviso.

Viewed as a whole, the revivalists are a remarkably diverse lot. Some were Dominicans, others Franciscans; Benedict was a layman, Guala was a bishop; Leo came from the highest aristocracy, John was the son of a notary; Gerard was politically pro-imperial, Leo was an enemy of the empire; Jacopino was highly educated, John (reportedly) and Benedict were 'of little learning'. It is only with great difficulty, then, that one can argue that they carried into the revival some hidden political or economic agenda. What alone united them seems to have been, above all, their common style of preaching and their focus on reconciliation and peace-making. That union of preaching and peace-making is best exemplified in the activity of the Dominican John of Vicenza.

The Great Prophet of Bologna

IN the late summer of 1231, Bishop Henry della Fratta of Bologna imposed an interdict on the government and inhabitants of his city. It was a drastic but not an unusual event in early thirteenth-century Italy. He had long been angered by the city's encroachments on the rights of jurisdiction that his bishopric had inherited from the old imperial administration, and by its resistance to his collection of tithes. Such communal encroachment on episcopal jurisdiction had been going on in the rural mountain-settlements to the south and in the fertile agricultural valley-lands to the north since he had taken up his office. These rights of jurisdiction seem to have gone back to the early twelfth century. Bishop Gerard Ariosti had received a privilege from the Emperor Otto IV confirming them in 1210. Gerard had been an inept pastor, and finally, after many corrections by the pope, he retired from office in 1213.[1] The new bishop, Henry, was of a more energetic sort. In 1215 Pope Innocent III summoned him, along with the other prelates of western Christendom, to attend his great reform council at the Lateran in Rome. While Henry was at Rome, where he must have heard Innocent's projects for promoting the bishop as the leading preacher and teacher of his diocese, his subjects back in Bologna were at work enhancing the role of the commune.

The Bolognese took advantage of Henry's absence at the Great Council to haul for trial before the municipal tribunal a resident of San Giovanni in Persiceto, an area over which the bishop traditionally exercised rights of justice.[2] They gaoled him and refused to turn him over to the bishop's court. From Rome, Henry struck back with the Spiritual Sword. The commune remained unmoved by the bishop's sanctions, and he had to wait until after his return in 1217 to entrust the dispute to the arbitration

---

[1] On these events, see Alfred Hessel, *Storia della città di Bologna dal 1116 al 1280*, ed. and trans. Gina Fasoli (Bologna, 1975), 208–9.

[2] On Bologna in this period, see ibid. 208–9. His narrative depends in part on letters contained in a unique 15th-cent. copy of the *Privilegia Episcopatus Bononiae* (Bologna, Archivio Notarile MS file 43, n. 99), described by him, p. 19, n. 154.

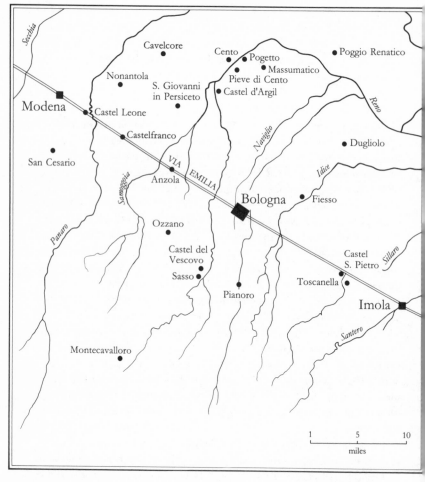

MAP 2. The *Contado* of Bologna. After Alfred Hessel, Storia della città di Bologna dal 1116 al 1280, ed. and trans. Gina Fasoli, (Bologna, 1975), p. xxix. For northern Italy, see Map 1.

of two masters of the University of Bologna.[3] The Bolognese masters were leading the revival of the Roman Law, with its exaltation of the powers of the emperor and his administrators, of which Henry was one. But they also knew that their residence in Bologna depended on the good will of a city government that was aggressively asserting its independence of the empire. Even if they were on Henry's side in their hearts, prudence seems

[3] Savioli, i. doc. 448; 449.

to have prevailed. They seem never to have rendered a decision. When help came to Henry, it came in the person of the emperor himself. In January of 1221, Frederick II confirmed Henry's ancient juridical rights and ordered the imperial legate, Conrad of Metz, to prevail upon the commune to respect them.[4] Henry remembered his friends: when war broke out in 1226 between the emperor and the Lombard League supported by Bologna, the bishop sided with Frederick.

During the first year of the war, the commercial and neighbourhood corporations that represented the Bolognese popular party rioted in the streets and successfully demanded a role in the city government. After all, they made up the bulk of the municipal militia and were dying in the commune's war against the emperor; it was only fair that they have a say in the affairs of the city. Bologna's popular revolution was a little late in coming. Similar popular revolts had occurred at Milan in 1207, at Vicenza in 1217, at Verona in 1229.[5] The communal governments against which they revolted were, at best, a concession granted begrudgingly to the cities of the Lombard League by the Emperor Frederick Barbarossa at the Peace of Constance in 1183. From the imperial point of view, the popular party and its organizations had no legal rights in the government and the commune had no right to grant them any. Nevertheless, they enacted their own laws and enforced the peace among their members. In each city the bishop, commune, and populars jockeyed for position in governing.

In the 3 years after the revolt of 1228, the corporations of the Bolognese popular party moved energetically to expand their role in the commune.[6] They quickly became the dominant force in the government, and city statutes reflected this.[7] The popular party's entry into Bolognese politics did give a political voice to parts of the population previously excluded. It simultaneously created new tensions, because their precise legal status was initially very ill defined. The popular party's principal vehicles of expression were its two associations, the Società delle Arti and the Società delle Armi. These two societies joined to create the

---

[4] RI, i. 272, letters 1220 and 1221.

[5] On these revolutions, see J. K. Hyde, *Society and Politics in Medieval Italy: The Evolution of the Civic Life, 1000–1350* (New Studies in Medieval History; London, 1973), 104–18.

[6] See Gina Fasoli, 'Le compagnie delle arti a Bologna fino al principio del secolo XV', *L'Archiginnasio*, 30 (1935), esp. 237–9.

[7] The tension created by these unstable political conditions seems to have been resolved in 1233 by a formal recognition of the *popolo*'s role in government; see ead., 'Le compagnie delle armi a Bologna', *L'Archiginnasio*, 28 (1933), 179–80. John had close connections with the Società delle Armi, and those bodies reaffirmed their unity by retaking their oaths in 1233; for his possible involvement in this reorganization, see Augusto Gaudenzi, 'Gli statuti delle Società delle Armi del Popolo di Bologna', *Bollettino dell'Istituto Storico Italiano*, 8 (1889), 201.

corporation of the *popolo*, 'the people,' sometime before 1231. In that year the *popolo* itself received direct voice in civic affairs through their governing board of 'ancients'. These three bodies' functions overlapped with those of the traditional councils of the commune, so conflict was inevitable.[8]

Remarkably enough, until the year 1230 Henry had remained on good terms with the popular-dominated city in spite of his pro-imperial stance.[9] Perhaps it helped that from 1227 to 1230 the Emperor Frederick was himself under papal excommunication for failing to keep his crusading vow and then for going on crusade without the pope's sanction. Churchman and commune could find a common enemy in the excommunicate emperor. In 1230 Frederick was reconciled with the pope at the Peace of San Germano. Henry availed himself of both papal and imperial support in his struggle for the bishopric's traditional rights; from then on his relations with the commune went from bad to worse.[10] The city government now represented a greater part of the citizenry than ever before and could rely on an ever greater base of support in its conflict with the bishop.

As Henry pressed his rights with greater and greater vigour, the attitude of the city turned from sullen acquiescence to outright defiance.[11] The new *podestà*, Frederick of Lavellongo, reacted quickly, sending his judges into areas under the bishop's jurisdiction and demanding a profession of homage from one hundred of the bishop's men. The commune then exacted an oath from all lay citizens in which they swore to refuse all assistance to the bishop's administration of justice and collection of tithes. Whether they were from the populars or the traditional ruling class, all could now bury their differences and present a united front against the pretensions of a bishop who until recently had been exhibiting pro-imperial sympathies.

Resistance to tithes and oaths of defiance were bad enough, but for Henry the final straw came when the commune began actively to investigate crimes at Dugliolo, an area whose jurisdiction belonged to him.

---

[8] On the *popolo*'s entry into the communal government, see Gina Fasoli, 'Oligarchia e ceti popolari nelle città padane fra il XIII e il XIV secolo', in Reinhard Elze and Gina Fasoli (ed.), *Aristocrazia cittadina e ceti popolari nel tardo medioevo in Italia e Germania* (Annali dell'Istituto Storico Italo-Germanico, 13; Bologna, 1984), esp. 23–4.

[9] Cf. Savioli, i, docs. 555, 580, 581.

[10] Hessel, *Storia della città di Bologna*, 209.

[11] On similar conflicts in the region, see G. Salvemini, 'La lotta fra stato e chiesa nei comuni italiani durante il secolo XIII', in *Studi storici* (Florence, 1901), 39–90; on the Church in communal statutes, see S. Pivano, *Stato e chiesa negli statuti comunali italiani* (Turin, 1904).

On 27 January 1231, Pope Gregory IX reviewed these and other in-
fractions stretching back to the events at San Giovanni in 1215 in a letter
that commissioned Palmerio, a canon of the Bolognese collegiate church
of Santa Trinità, to intervene on the bishop's behalf.[12] Gregory reserved
to himself, at least for the meantime, the power to excommunicate or
interdict the commune, if that became necessary. Palmerio's efforts came
to nothing.[13] Most likely the newly united commune saw no reason to
compromise.

Henry was reduced to inactivity. The energetic *podestà*, on the other
hand, set about seizing the bishop's castles in the ecclesiastical territories
of San Giovanni in Persiceto, Anzola, Dugliolo, Castello del Vescovo,
Massumatico, and Montecavalloro. With the pope having reserved the
right of excommunication to himself, Henry was left to fume for 6
months. Then, in September 1231, he retaliated against the commune
with an interdict and in February of 1232 abandoned the city for Reggio,
where he spent the winter, spring, and summer.[14] His interdict remained
in force for about 10 months.[15] The *podestà*, Frederick, probably sensing
that his policy had wide support, refused to budge.

In 1232 Rainier Zeno of Venice became *podestà* of Bologna.[16] Papal
legates immediately approached the new *podestà* demanding the return
of the confiscated castles. Rainier agreed, but only on condition that
municipal judges (*praetores*) be allowed entry into those districts to administer justice in the name of the commune. Since the real dispute was
over jurisdiction, not over castles, this was no compromise at all. The
pope responded in anger, ordering bishops Grazia of Parma and Guidotto
of Mantua to promulgate a papal excommunication against the government and city of Bologna. Students had to leave Bologna, and the
activities of its famous university were suspended.[17] For a city whose
chief claim to glory was its famous university, this suspension was a
threatening portent. Should it endure too long, the masters and scholars
might well decide to move elsewhere. Bolognese honour and reputation

---

[12] Auvray, i. 346–7, letter 534, text in RI, 6833. This letter's description of the
commune's invasion of episcopal rights as sacrilege was later included in X 2. 2. 16.

[13] RI ii. 1200, letter 6894; Savioli, iii. 119.

[14] As seen in *BOFM* ii. 71.

[15] Rampona, 99: 'Item lo dicto anno circa calende de' setembre naque discordia tra el
vescovo de Bologna e'l comun de quella, in tanto che la città sté interdicta circa el spacio di
x misi.'

[16] Matteo Griffoni, *Memoriale Historicum de Rebus Boniensium*, ed. Lodovico Frati and
Albano Sorbelli (RIS² 8:2; Città di Castello, 1902), 9.

[17] The history of this quarrel is known from the 2 Jun. 1232 letter of Pope Gregory IX,
edited in Savioli, iii. 119. See also Sutter, 64–6.

were at stake. The danger was real. In the early 1220s disgruntled scholars had attempted to found a university at Padua, and other attempts had been made in previous years. There was always the danger they would try again.

On 6 June 1232 the parties reached an agreement on the tithe question, and the papal excommunication was lifted. In compensation for the loss of the tithes, the bishop received control of the communal castles in Cento and Pieve di Cento, two settlements flanking the River Reno north of the city. They left the quarrel over jurisdiction unresolved; perhaps on all sides there was a growing feeling that some reconciliation had become necessary. By 20 October 1232 Bishop Henry had re-entered his city amid outbursts of joy.[18] The public burning of sorcerers' manuals amid general rejoicing, which chroniclers record in that year, suggests a renewed co-operation between the secular and ecclesiastical authorities.[19] Nevertheless, the city had made peace with its bishop only under pressure, and the commune had offered no recognition of episcopal jurisdiction in the *contado*. Prospects for lasting peace remained dim.

This jurisdictional dispute had occurred against the background of an economic crisis triggered by bad weather and poor harvests. Bologna, the largest city of the Po valley and a centre of the grain trade, was especially hard hit. The failure of the grain harvest in both 1232 and 1233 drove up the price of bread. During the winter of the latter year, rioting crowds seized the city's millers, tied their hands behind their backs and flogged them, naked, through the streets in retribution for what were considered exorbitantly high flour prices.[20] As in all times of want, passions against usurers ran high; rioting and looting of pawnshops broke out later in the spring.[21]

Ominous threats from without overshadowed the city as well. No longer united by the need to oppose the emperor and demand the right of self-government, the north Italian cities had broken into factions during

[18] Rampona, 99: '[E]t a sei dì de luglio fu facta la concordia e lo vescovo Henrigo tornò a Bologna cum grande festa et allegreza de cittadini.' Ignorance of this, at that time unpublished, text led Sutter, 65, to believe that when John of Vicenza arrived the university was still closed and the interdict still in force; an error repeated by Daniela Gatti, 'Religiosità populare e movimento di pace nel'Emilia del Secolo XIII, in *Itinerari Storici: Il medioevo in Emilia* (Carpi, 1980), 103.

[19] Griffoni, *Memoriale Historicum*, 9; also in Borselli, *Cron.*, 22 and Varignana, 101.

[20] Borselli, *Cron.*, 22; also briefly reported in Rampona, 101, Varignana, 101, and Griffoni, *Memoriale Historicum*, 9.

[21] Hessel, *Storia della città di Bologna*, 156–7, sees the period from 1230 to 1250 as a period of economic stagnation and the beginning of Bologna's decline as the great power of Emilia.

the first decades of the 1200s. It was the inevitable result of each city's struggle to extend its control into the hinterland—each city eventually bumped up against the next city's sphere of influence. One then allied with other cities whose expansion had been blocked in the same way. The result was a chequer-board of alliances across north Italy.

War could be the immediate result. For 3 years, from 1226 to 1229, Bologna embarked on a series of bloody campaigns against the castles and troops of its neighbour Modena. During this conflict Modena received military support from Parma and Cremona, while Bologna's erstwhile ally, Reggio (Modena's neighbour to the west), remained neutral. During the very unsuccessful campaign of 1228, an outbreak of plague at San Cesario in the Modenese *contado* forced the Bolognese army to withdraw. Popular elements in the municipal militia revolted, demanding a role in the government and forcing the city to sue for peace.[22] Bishop Henry, still on moderately good terms with the commune, even served on the commission representing Bologna in peace negotiations with the Modenese.

On 10 December 1229, a truce negotiated by the Dominican preacher Guala of Bergamo took effect.[23] It was intended to last for 9 years. It lasted only until 1234, when the Bolognese again took up arms against Modena.[24] Although there is no evidence that the Bolognese army engaged the Modenese before that year, preparations for war against Modena and its western allies were certainly already under way in late 1232, when the Bolognese refortified Castelfranco, their western stronghold defending the Emilian Way. They not only restored the castle's walls but also provided them with new towers.[25] The two armies, it would appear, stood facing each other across the refortified frontier, each waiting for the other to make a provocative move.

To the food shortages, riots, military threats, and political divisions, the heavens themselves added a disquieting omen: in July of 1232 two planets underwent conjunction in the night sky over Bologna, an event recorded with foreboding by frightened chroniclers.

---

[22] Ibid. 174. This riot and the occupation of the *palazzo comunale* is described in Pietro Cantinelli, *Chronicon {AA 1228–1306}*, ed. Francesco Torraca (RIS² 28:2), where it is misdated to 1231.

[23] Bolognetti, 98.

[24] The truce was not broken in 1229, as Villola, 98, reports, although tensions must have been running high.

[25] Borselli, *Cron.*, 22; reported also in Varignana, 101, and Griffoni, *Memoriale Historicum*, 9. For political and military action from 1227 to 1234, see Hessel, *Storia della città de Bologna*, 104–10.

FRIAR JOHN OF VICENZA

Against this background, in the spring of 1233 the preacher John of Vicenza arrived in Bologna and the city soon hailed him as a great prophet and mediator.[26] These were titles whose reality he still had to prove. Whatever the praises they bestowed on him in the early spring, the commune did not commission him to help resolve their jurisdictional dispute with the bishop until 19 April, some 2 months after his arrival.[27] The exact date of his arrival is unknown. It is possible that John reached Bologna during Lent (9 February to 24 March in 1233) since the Bolognese council fathers would tell the Dominicans who had assembled for their order's general chapter (25 May) that John had fruitfully sowed the Word of the Lord among them at that time.[28] His arrival probably dates after the end of 1232, since at that time it seems he was still present among the Dominican and Franciscan revivalists preaching at nearby Modena.[29] It is likely that during that winter of 1233 other revivalists were spreading John's reputation as a prophet and mediator, and the Bolognese may themselves have invited him to their city.[30] They had many reasons to seek the advice and consolation of a messenger of God in the early months of 1233. Whether John was invited by the commune or not, his first sermons had to establish his reputation as a divinely inspired preacher. He adapted his message to Bologna's anxieties and needs. Soon after his arrival John's impact on his hearers was such that everyone accepted his words as those of Christ himself.[31] He had heard the spoken, and sensed the unspoken, needs of the city.

[26] Villola, 101: 'Iohannes de ordine fratrum Predicatorum venit primo civitatem Bononie, et dicebatur magnus profeta.' See also Borselli, *Cron.*, 22, and Rampona, 102. His detractor Guido Bonatti, *De Astronomia Tractatus X*, ed. Nicolaus Prukner (Basel, 1550), col. 210 reports: '[R]eputabatur sanctus quasi ab omnibus Italis, qui confitebantur ecclesiam Romanam; mihi autem videbatur quod esset hypocrita.'

[27] For this reason I doubt Sutter's contention that John came to Bologna specifically to arbitrate this dispute.

[28] Borselli, *Cron.*, 22: '[I]n quadragesima preterita divinum verbum seminaverat in populo.' *VF*, 138–9, also records John's appearance at the chapter but not his preaching during Lent.

[29] *Ann. Vet. Mut.*, col. 60: 'Eodem Anno Predicatores Fratrum Minorum, et S. Dominini [*sic*] crescere ceperunt cum Confalonibus, & Crucibus.'

[30] Contrary to Sutter, 63–4, who follows Hessel, *Storia della città di Bologna*, 215, it is unlikely that Jordan of Saxony, the Dominican master general, invited John to Bologna. Returning from France, Jordan spent Lent of 1233 in Reggio organizing the foundation of the Dominican convent there and probably arrived in Bologna only in May for the general chapter. See Marguerite Aron, *St Dominic's Successor* (London, 1955), 169–70.

[31] Borselli, *Cron.*, 22: '[T]aliter predicavit et cum tanta gratia verbum Dei seminavit, ut cives, milites, rustici et omnes quos ad divinorum mandatorum obedientiam traxerat, ita

First, John addressed the economic crisis. He preached against usurers and others who exploited the People of God during times of shortages. His words must have touched a nerve in his hearers, since this homily provoked a riot during which crowds burned the house and records of one of Bologna's chief usurers, Pasquale Landolfo. The poor man escaped lynching only by fleeing the city.[32] John called for the freeing of paupers imprisoned for debt and for the pardoning of all liabilities to creditors.[33] Such preaching, of itself, won him immediate popularity, at least with debtors and their relatives. He preached, too, against vanities, such as women's extravagance in dress, and this must have evoked similar positive feelings among those too poor to accumulate such luxuries. The banning of extravagant displays of wealth, like the cancelling of debts, removed potential causes of hostility in the divided community. John did not restrict his work to sermons. He added a personal touch of compassion to his ministry. He went into the homes of those who could not attend his sermons. He listened to their troubles and spoke words of consolation. He visited the sick and prayed over them. It was not long before the stories began to circulate that he had miraculously healed many with the Sign of the Cross.[34]

John also addressed the military causes of the city's anxiety. One story reported by Salimbene, which sheds light on this aspect of John's rise to power, may be placed during his early Lenten preaching, probably during the period of Modena and Bologna's refortification of their frontier.

---

eum venerabantur ut quecumque dicebat tanquam a Christo fuissent dicta, susciperent devotissime.' And in Rampona, 102 (Borselli's source): '[C]he per tale modo predegava al puovolo, che tucti li citadini, contadini et destretuali, de Bologna in tanto li credevano che lo seguiano predicatione et commandamenti.'

[32] The oldest chronicles (early 14th cent.) do not agree on the date of this riot: Villola places it in 1232, Bolognetti in 1233. The two late 14th-cent. chronicles follow the earlier according to their usual pattern: Rampona follows Villola, Varignana follows Bolognetti. Griffoni (written 1404–26) follows Villola as usual. The significant chronicle here is Borselli (late 1400s). He usually follows Rampona, but here he follows the other tradition, putting the event in 1233. Furthermore, he explicitly ascribes the riot to John's preaching. With good reason, Sutter and Vauchez accept this report. Borselli, a resident of the Dominican house in Bologna, is especially well-informed about John and the Dominicans. He, for example, corrects the error in all other sources that the translation of Dominic's body to 'una archa che era molto ben scholpida' took place on 23 May 1233. The translation to the sculpted *arca* (that of Nicola Pisano) occurred in 1260. The *arca* of 1233 was, as Borselli says, 'marmoream sine figuris sculptam'. If trustworthy in small things, he is so also in greater ones.

[33] Borselli, *Cron.*, 22: 'Propter debita incarceratos liberos fecit, aut creditoribus soluta peccunia, aut ab eis totaliter dimissa.' More vaguely, Rampona, 102: '[E] fé relasare tucti li presonieri delle carcere de Bologna.'

[34] Borselli, *CMG*, fo. 21[r], edited in App.

During a sermon delivered at Parma, John's associate, Gerard of Modena, fell into a trance and, on regaining his senses, announced that he had received a vision of John preaching to the Bolognese on the banks of the Reno. Parma was allied with Modena in that city's war with Bologna.[35] The theme of John's homily, Gerard revealed, was that the Bolognese hearers were blessed since they were God's children.

Spies at Bologna not only found the vision to be accurate, but also reported that John had himself received a vision—one revealing the topic of Gerard's sermon in Parma.[36] Gerard's news that Parma's Bolognese enemies were God's children surely helped to open the channels of peace-making between the two cities. At this early date John was already laying the groundwork for projects far more extensive than those in Bologna alone.

It would be fascinating to know something of the vision which John revealed to the Bolognese. In any case the reaction of John's hearers in Bologna must have mirrored that of Gerard's at Parma. The banks of the Reno were the perfect place to address the issues of war and peace. Only a short distance up the Emilian Way, west of the Reno where John was preaching, lay Castelfranco, the castle already being refortified for renewed war against Modena and its Parmese allies. The citizens of the economically strapped city, seeing little value in another costly war of expansion, must have welcomed a divine message offering the possibility of peace with their enemies. Not surprisingly, as his career developed, John found his staunchest supporters among the peace-faction in Bologna—the same Società delle Armi whose revolt had forced the truce with Modena in 1229. They were, for the most part, the same tradesmen and artisans who had been especially hurt by the shortages and predations of usurers.

Having addressed his hearers' anxieties and received supernatural support for his activity through his visions and healings, John had then to create the mechanisms of publicity that would confirm and propagate this growing reputation. Following his initial successes, John began to

[35] Although Salimbene places Gerard's vision in Parma, Gerard also preached in Modena itself, see *Ann. Vet. Mut.*, col. 60 (1233): 'Eodem Anno factae fuerunt paces Mutinensium mediante Fratre Gerardo Ordinis Minorum, et omnes quacumque de causa a Communi Mutinae banniti reversi sunt, praeter quinque. Eodem Anno multae praedicationes factae fuerunt.'

[36] Salimbene, 76–7.

organize displays of support by his followers, thereby overawing sceptics and winning even greater notice.[37] The Bolognese formed themselves into processions and marched to his sermons with crosses and banners. Among the organizations present were the companies of the Società delle Armi and, reportedly, other bands of tough young men armed with quarter-staves.[38] John's spiritual power was taking on a more noticeably martial flavour. Along with the processions going to hear his sermons, John amassed others in which the participants carried banners and candles while singing the praises of the name of Jesus.[39] As the revival intensified, John's marching followers became almost a daily sight in the streets.[40]

By late April, John had spread his message from Bologna to nearby cities. On 24 April the people of Faenza (Bologna's eastern ally) marched with their banners some 15 miles through the *contado* of hostile Imola to Castel San Pietro in the *contado* of Bologna to hear John speak. They even hoped, with luck, to see him work a miracle or two.[41] Guido Bonatti, who observed these events from Forlì (Faenza's enemy to the east), reported with distaste that John's swarms of admirers were now becoming frenzied at his appearance, hurdling over one another to rip threads from his habit as relics.[42] There was no question that the preacher had become

---

[37] Although far from Bologna, Matthew Paris, *Chronica Maiora*, ed. Henry Richard Luard (Rolls Series, 57, 7 vols.; London, 1872–83), iii. 287, emphasizes the characteristic use of processions by Alleluia preachers: 'In tantam nobilitatem, ne dicam arrogantiam, elevabantur Praedicatores et Minores, qui spontane paupertatem cum humilitate elegerunt, ut recipi curarent in cenobiis et civitatibus in processiones sollempnes in vexillis, cereis accensis et in disposicione vestimentis festivis indutorum.' Matthew's suggestion that the friars' voluntary poverty helped their popularity is attractive, but I find little to confirm it directly.

[38] On these processions and the Società delle Armi, see Borselli, *Cron.*, 22, and Rampona, 102. Bonatti, *De Astronomia Tractatus*, col. 211, tells us of the armed bully-boys. He declares that they roughed up hecklers and that John enjoyed the show. Bonatti was an eye-witness but a prejudiced one. Still, his accusation has a ring of truth to it.

[39] Rampona, 102.

[40] We do not know how frequently John preached, but an indulgence granted by Pope Gregory IX to John's hearers in the Veneto by a letter of 13 Jul. 1233, *BOP*, i. 57, Letter 88 (Auvray, i, col. 813, letter 1461), speaks of solemn preachings 'tribus diebus in hebdomada', a not unlikely number.

[41] *Chronicon Faventinum Magistri Tolosani {AA 20 av. c.-1236}*, ed. Giuseppe Rossini (RIS² 28; Bologna, 1939), 159: 'Post gloriosum Virginis partum MCCXXXIII, die VI exeunte mense aprilis, iverunt Faventini omnes cum pueris et puellis, senes cum iunioribus atque mulieribus apud Castrum Sancti Petri de Bononia, tam de civitate quam de districtu, cum vexillis omnibus, cruces desuper portantibus, ad predicationem fratris Iohannis, qui est de ordine Predicatorum, qui et Jesus Christus multa per eum mirabilia operatus est.'

[42] Bonatti, *De Astronomia*, col. 210.

a force to be reckoned with. He had arrived; he had earned his title, the Prophet of Bologna.[43]

BOLOGNA PACIFIED

By mid-April John was at the zenith of his influence in Bologna. His activities extended beyond organizing processions and preaching. He entered the political sphere—mediating in disputes and bringing about reconciliations.[44] John's first reconciliations were undoubtedly between individuals and families. In the only one for which the issue at stake is known, John mediated between one Armanno of Porta Nuova and the widow and heirs of Gandulf of Gisso. The peace instruments for this property dispute are found among the statutes of Bologna.[45] There were many other private arbitrations, both in the city and in the *contado*, if John's work paralleled that of Gerard at Parma. By 19 April, John's reputation as a peace-maker received its pre-eminent recognition. On that date the bishop and the commune formally entrusted the arbitration of their jurisdictional dispute to him.[46]

At the time John received his commission, Bologna was nearly 3 weeks into Paschal Time, and the revivalist's processions were singing the Alleluias that gave the Great Devotion its name. Then, during the solemn fast that marked the Vigil of Pentecost on 14 May, John organized an enormous penance-procession in which all, or the greater part, of the citizens of Bologna participated. It was a suitable occasion, the preparation for Christ's sending of the Holy Spirit on the Apostles 'to teach them all things'.[47] The Bolognese walked barefoot and wailed aloud. This procession was part of the preparations for John's most celebrated sermon of the Bologna campaign, that to the council and governors of the city on the following Monday, 16 May.

---

[43] Lodovico Gatto, 'Il sentimento cittadino nella *Chronica* di Salimbene', in *La coscienza cittadina nei comuni italiani del duecento, 11–14 ottobre 1970* (Convegni del Centro di Studi sulla Spiritualità Medievale, 11; Todi, 1972), 370–2, suggests that John's public demonstrations nurtured Bolognese communal identity.

[44] Explicitly in Villola, 101: '[E]t propter suas predicationes fate fuerunt multe paces.' See also Borselli, *Cron.*, 22 and Rampona, 102.

[45] Bol. Stat., i. 448–9.

[46] Text found in Savioli, iii. 123–5, edited from Bologna, Archivio di Stato MS: *Registro Nuovo*, fo. 353ᵛ.

[47] John 13: 13.

On that day, convoked by the ringing of the bell on the Palazzo Comunale and the dispatching of criers throughout the city, the Bolognese and their leaders came to hear John address the council of the city.[48] Most likely there were few in the city who needed to be reminded of what was to occur that day. The Holy Spirit was surely upon the man of God. During this address a luminous cross appeared, visible to all, on John's forehead, and his hearers burst into tears at the sweetness of his words.[49] .

This was John's finest hour in Bologna. It is not impossible that on this occasion John received the commune's statute books for revision.[50] Following the practice of other preachers of the revival, John had not only reconciled the factions of the city but also demanded that he be given plenary powers to rewrite the city laws. How he found time to do this in the midst of his preaching and processions is hard to tell. Most likely he worked with several other preachers, adding, subtracting, and editing as appeared necessary. If their revision followed the better-documented pattern of Gerard at Parma, the major focus of the reform was to create means for arbitrating disputes and to institutionalize the reconciliations that John had helped bring about.

Unfortunately, almost nothing of John's work remains in the extant 1245 version of the Bologna statutes. The little that does remain is suggestive. He suppressed all oaths made by members of factions in Bologna and its district.[51] This statute was the centre-piece of John's programme of reconciliation, since members of such factions were

---

[48] Thus in accord with the norms for convoking the council of the city.

[49] Borselli, *Cron.*, 22: 'Super ipsum, existentem in consilio civitatis et consiliarios admonentem ad bene gubernandam civitatem, apparuit signum crucis omnibus ibi existentibus visibile. Ex dulcedine verborum eius multi plorabant.' Also in Rampona and Bolognetti, 102 (who give the date 'adì 16 de mazo'), and in Varignana, 101.

[50] As implied by the report of Borselli, *Cron.*, 22. Although no historian has made this connection, the artist Vittorio Bigari (d. 1776) did. His fresco in San Domenico in Bologna shows John, cross over his head, delivering the statutes to the city fathers. The inscription reads: 'Bononiae magister b. Joanni Schio urbis Statuta reformanda tradit, A. MCCXXIII.' Whether the error in the date is a slip of the painter's hand or of his informant's memory is unknown. Rampona, 102, and Borselli, *Cron.*, 22, describe the revision of the statutes.

[51] Found in Bol. Stat., ii. 262: 'De coniurationibus non faciendis Rubrica. Statuimus quod de cetero jn civitate vel districtu non fiant coniurationes vel promissiones vel sacramenta vel talia similia occasione parcium aliquarum jn civitate bon. vel districtu; et si quis contrafecerit, solvat nomine banni CC libras bononinorum. Item statuimus quod si aliqua facta fuerint inter nos aliquis [*sic*] jn civitate bon. vel districtu, quod omnia dissolvantur, et quod omnes ad invicem se absolvant, et ego fr. Johannes hec omnia casso et pronuncio nulla.' The reading of 'vel' for 'aliquis' suggested by Sutter, 76, seems a sound correction of a misread manuscript abbreviation.

pledged to protect and avenge each other against their enemies. The cessation of these oaths was undoubtedly the high point of his sermon. He went on to lecture the city fathers on how they could avoid the internal and external wars and division that plagued their city. Then, most likely as the final flourish of his address, John also ordered the release of those Bolognese imprisoned during feuds and factional strife.[52] The prisons were thrown open, the exiled returned, peace and concord reigned. Friar John was triumphant.

By the end of April news of John's success in Bologna had reached the ears of Pope Gregory IX. On 28 April, the papal chancery composed a letter to the preaching friar, congratulating him on his gift of miracle-working and urging him to accept the office of papal mediator between the warring cities of Florence and Siena.[53] The pope explicitly refrained from commanding John to accept: it would have been wrong, he said, to give commands to someone so obviously inspired by the Holy Spirit.[54] The following day the pope dispatched a second letter, this time to the commune of Bologna, ordering them not to obstruct John's freedom of movement.[55] With the letter to the commune went another, containing a blessing for the Dominican general chapter that was to open at Bologna on 25 May.[56] The three letters probably arrived in Bologna some ten days later, just about the time of John's penance procession and his address to the council of the city.

In mid-May, when the pope's letters arrived, John had no intention of abandoning Bologna for Tuscany. During the night of 23–4 May he was present for the translation of Dominic's body (which signalled the beginning of his canonization process) to a plain but noble new tomb.[57] This translation was above all the fruit of John's own preaching. Stephen

---

[52] Bonatti, *De Astronomia*, col. 210, says he ordered the release of one Lawrence, a knight condemned to death by the *podestà* for killing the son of a neighbour. Bonatti saw in this John's disdain for the law. The release was surely part of the terms of reconciliation in a blood feud.

[53] Had John gone to Florence he might not have received a cordial welcome, at least if we believe what Salimbene, 77–8, tells us. According to him, the Florentines greeted the reports of John's miracles and successes with scorn and said that, if it were true that John was raising the dead, he would find no welcome in Florence since it was already overpopulated.

[54] *BOP*, i. 48, letter 73 (Auvray, i. 713, letter 1270).

[55] *BOP*, i. 48–9, letter 74 (Auvray, i. 713, letter 1268).

[56] *BOP*, i. 49, letter 75 (Auvray, i. col. 714, letter 1271).

[57] His presence at the translation is recorded by Borselli, *Cron.*, 22, and *CMG*, fo. 22ʳ. On the translation, see also CCB, 102–3.

of Spain, the Dominican prior of Bologna, reported during the canonization process that in the spring of 1233 John had been putting it out in his sermons that he had been privy to a vision revealing the founder's holiness and power of intercession. John's preaching had almost singlehandedly created a new saint. Previously the Dominicans who guarded the tomb at San Nicola delle Vigne had discouraged any cult. The early Dominicans, unlike their Franciscan brothers, had shown little veneration for their founder. He had been left under the feet of the brethren as he had requested, in a grave beneath the choir of the church. Before 1233 no candles burned at the tomb and no crowds came to pray.[58] It seems that the Dominicans rebuffed pilgrims to the tomb as an intruding nuisance.

The deputation opened Dominic's grave, which had long been exposed to the elements because of construction, privately and at night, lest the discovery of a putrefied body scandalize the people. A large group was present: the papal legate Archbishop Theodoric of Ravenna, Bishop Henry of Bologna, Jordan the master of the Dominican order, John himself, many secular notables, and about 300 Dominicans. Had the body been corrupt, it would have been difficult to keep it a secret. There was no need for anxiety. When the tomb was opened, a sweet odour filled the church and clung to the hands of all who handled the relics.[59] John's visions had been miraculously validated. At this, the best-remembered incident of John's career, he must have outshone even the high prelates present.[60]

John responded to the pope's request that he come to Tuscany soon after receiving the letter, since on 26 May the pope wrote him a reply, rejoicing over John's successes and miracles.[61] In this letter there is no suggestion of sending the friar to Tuscany. Perhaps the pope temporarily abandoned the idea after hearing about the importance of John's work in Bologna. In the eyes of the Bolognese, John was now well-nigh indispensable. To the Dominicans assembled at San Nicola delle Vigne for

---

[58] So Salimbene, 72.

[59] For the earliest description of the translation, see Jordan of Saxony, *Liber de Principiis Ordinis Praedicatorum*, ed. H.-C. Sheeben (MOFPH 16: MHD 2; Rome, 1935), 86–8. He does not mention John's presence.

[60] A surely apocryphal story in Th. Cant., 424, symbolizes John's importance in the translation. He says John had been positioned at the feet of the coffin and the papal legate at the head. When the coffin was opened Dominic was found to have turned so that John was at the head. Instead of taking the body out, they closed the coffin and the friar and bishop changed positions. At the second opening, Dominic had reversed direction again. The defeated bishop then allowed the body to be removed, with John holding the head end. A carving on Dominic's tomb immortalizes the story.

[61] *BOP*, i. 51, letter 78 (Auvray, i. 751, letter 1339).

their general chapter, the commune sent some of their most important
and learned officials to beg that John not be moved from the city but be
allowed to stay on and bring his work to completion. They feared the
fruits of his peace-making and preaching would be lost if he left. Their
many arguments were of no avail. Jordan, the leader of the Dominicans,
perhaps influenced by the pope's earlier request that John move on to
Tuscany, politely refused. 'One sows and another reaps in the vineyard of
the Lord' was the gist of his reply. Jordan did, however, promise to take
counsel with the assembled friars and provide for the city's needs.[62]
When the pope's letter of 26 May, which praised John and made no
further mention of Tuscany, arrived, Jordan must have acceded to the
wishes of the city, since on 20 June he himself witnessed the document
by which John officially completed his peace-making between the city
and commune.[63] The pope, however, had not yet abandoned hope of
inducing John to come south.

At the end of the same week, on Saturday, 28 May, John left Bologna for
'Lombardy', which probably means the lands of Bologna's enemies to the
west, Modena, Parma, and Cremona. There he preached in the midst of
hostile forces assembled on the field of battle.[64] His preaching was so
effective that the armies dispersed and returned home.[65] He had postponed the long-threatening war between Bologna and its western neighbours for the time being. The triumphant preacher returned to Bologna

---

[62] The events at the chapter recorded in *VF*, 138–9, repeated by Borselli, *Cron.*, 22–3.

[63] See document in Savioli, iii. 132.

[64] See *Doppelchronik*, 164, a report which corresponds to Salimbene, 74: 'Magnam
predicationem inter Castrum-Leonem et Castrum-Francum tempore illo fecit'; repeated in
Memoriale Potestatum Regiensium, ed. Lodovico Antonio Muratori (RIS 8). col. 1108.

[65] Rampona, 103: 'Et adì 28 de mazo del dicto anno fra Zohanne andò in Lombardia e fé
fare de molte paxe de gran guerre che gli erano tra li comuni de le terre e fé partire li hosti
che erano a campo, et predeghò in li dicti hosti.' Repeated by Borselli, *Cron.*, 23, who adds:
'Obsidiones multorum circa castra et civitates abire precepit.' Bolognetti, 102 and
Varignana, 102, give reports almost identical to Rampona. Sutter, 91–2, rejects the dates
in these texts and links the events in them to John's peace-making campaign of early July in
the Veneto. He thinks that a peace campaign in late May would prevent John's return to
Bologna in time to promulgate his first arbitration-decree on 31 May. I see no reason to
displace these reports. John could easily have returned to Bologna after a weekend of
preaching at some not too distant place in 'Lombardy', such as Modena or even Parma. Nor
does this description of John dispersing troops in the field sound like any incident of the
Veneto Campaign. During that campaign the Bolognese sent forces against Modena to force
that city to join the Lombard cause. That expedition ended in a Bolognese victory, not a
truce mediated by a preaching friar.

and issued his arbitration between Bishop Henry and the commune during a solemn sermon in what is now the Piazza del Mercato. The peace instrument he drafted for this occasion has, unfortunately, been lost; its existence is known solely from its cancellation in John's final arbitration-decree. Nothing is recorded concerning its reception or the events that made its revision necessary.

On the other hand, the text of the final arbitration, which John promulgated on Monday, 20 June, still exists.[66] This document is almost completely in favour of the commune. In four places of disputed juris-diction (Castel d'Argile, Ozzano, Fiesso, and Montecavalloro), the bishop lost all rights of justice; in the other six (San Giovanni in Persiceto, Anzola, Castel del Vescovo, Poggetto, Massumatico, and Dugliolo), these were reduced to encompass only minor crimes. John reserved to the *podestà* all crimes touching public welfare. Finally, those who admin-istered the bishop's justice were to pledge fealty to the commune, and the commune received the right to place an inspector in areas of episcopal jurisdiction to supervise weights and measures, raise the militia, and ban rebels. Perhaps John's first arbitration had been too favourable to the bishop and thus not well received by the commune.[67] If the first had favoured the Church, the second was a major victory for the commune over the traditional jurisdictional rights of the bishop.[68]

John had accomplished his work at Bologna and he now looked to carry his revival further afield. He had not anticipated the difficulties of becoming the hero of the Bolognese. They now so idolized him that, when he decided to embark on a new preaching-campaign in the Veneto and the Veronese March at the end of June, he had to escape from the city at night and flee to Modena. From there, he and his supporter and friend, Bishop William of Modena, went by small boat down the Panaro to the Po and then on to Ferrara. From there he went north, past Rovigo. He

---

[66] Text in Savioli, iii. 128–30, edited from Bologna, Archivio di Stato MS: *Registro Nuovo*, fo. 354ᵛ. The text was also entered, in a defective form, in Bologna, Archivio di Stato MS; *Registro Grosso*, i, fo. 514ʳ⁻ᵛ. Savioli's heading errs in giving the date of this decree as 24 June. The document itself reads clearly: 'die lune undecima exeunte Junio'. Carlo Sigonio, *Historia de Rebus Bononiensibus Libri VIII* (Frankfurt, 1604), 107, makes the same error. Antonio Magrini, *Notizie di fra Giovanni da Schio per le nozze di Nanne Gozzadini e Maria Teresa Sarego* (Padua, 1841), 9 and 7, misconstrues this date as 11 June. 11 June 1233 was not a Monday.

[67] Sutter, 92.

[68] So too Hessel, *Storia della città di Bologna*, 209–10.

then entered the district controlled by Padua, passing under the great hilltop fortress at Monselice. Padua and the Veneto lay open to him.[69]

John returned to Bologna several times during the summer of his northern campaign—the city seems to have remained his base of operations—but, from June on, his sphere of activity was no longer a single city, however important; it had expanded to encompass almost all of north-eastern Italy. Unlike the Bologna campaign, which resembled those of other Alleluia preachers, his Veneto campaign was unique. No other revivalist of 1233 ever operated on such a vast scale.

---

[69] *Ann. Vet. Veron.*, 92: 'Et inde secedens de nocte clam Juit mutinam. Et postea cum episcopo mutinensi in una nauicula fugit ferrariam quia bononienses volebant eum libentissime habere, et postea transtulit se paduam.'

3
Peace Campaigns in the North

THE conflicts which John had confronted and resolved at Bologna were intense but local. The war between Bologna and Modena was already unpopular and the sides were clearly defined. In the north, John faced strife that extended far beyond the walls of a single city, and the alliances between those involved were often unclear and shifting. The Veneto of 1233 would have presented a trial to even the most skilled diplomat.

Although the politics of the Veneto and the Marches of Treviso and Verona had been exceedingly fluid in the early 1230s, by the opening of the year 1233 the communes and factions of the region had coalesced into two recognizable but still unstable groupings, one under the leadership of Azzo, the duke of Este, the other under that of the two da Romano brothers, Ezzelino and Alberico.[1] Independent of both parties was the commune of Padua, where the Benedictine Jordan Forzaté dominated local politics.[2] In 1233 Padua was nominally allied with Azzo's party. Contention between the two alliances had become especially fierce during 1232 and early 1233.[3]

The centre of da Romano power was Verona, dominated in this period by Ezzelino himself. His brother Alberico's power was concentrated in the March of Treviso. In both Vicenza and Verona, factions known as the Quatuorviginti and the Montecchi—the latter called after the residence of

---

[1] For a brief overview of the Veneto politics in this period, see J. K. Hyde, *Society and Politics in Medieval Italy: The Evolution of Civic Life, 1000–1350* (New Studies in Medieval History; London, 1973), 119–23, and for bibliography on the popular communes and the *signorie*, ibid. 210–11; on Ezzelino da Romano, see Gina Fasoli (ed.), *Studi ezzeliniani* (Rome, 1963), particularly, Raoul Manselli, 'Ezzelino da Romano nella politica italiana del secolo XIII'. For political developments leading up to Paquara, see Giovanni and Alvise da Schio, *Fra Giovanni da Vicenza a Paquara: 28 agosto 1233–1933* (Schio, 1933), but this adds little to Sutter, 94–6.

[2] On Jordan Forzaté in Padua, see the biography by Italo Rosa, *Il beato Giordano Forzaté, abbate e priore di San Benedetto in Padova, 1158–1248* (Bresseo, 1932), and, more recently, Antonio Rigon, 'Vescovi e ordini religiosi a Padova nel primo duecento', in *Storia e cultura a Padova nell'età di sant'Antonio* (Fonti e ricerche di storia ecclesiastica padovana, 16; Padua, 1985), 135–41.

[3] Parisio, 8; Maurisio, 28–9; Rolandino, 44; *Ann. Vet. Veron.*, 91. See also Sutter, 94–7.

Uguccione da Pilio, a powerful citizen of Vicenza—usually supported the da Romano.[4] Scholars generally consider a new alliance of the da Romano with Treviso and Count Guido of Vicenza as having strengthened their party in the early 1230s. This allowed it to dominate Conegliano, Treviso's weaker northern neighbor.

Traditionally, scholars have called the da Romano group the Imperial (or Ghibelline[5]) party, since the da Romano held Verona in the name of the Emperor Frederick.[6] Their shifting alliances will show that the da Romano's 'imperial' connection was more opportunistic than it was ideological. This group may be called the da Romano party for convenience, even though its different members had a variety of interests that did not always correspond with those of that family. Opposed to the da Romano was the loose collection of cities and magnates sometimes called the Guelfs or the Lombard party, after the league to which most belonged. For want of a better name, the latter is employed here. They were a diverse group. With the support of Azzo d'Este, Richard of San Bonifacio, and the da Camposampiero family, the commune of Mantua threatened Ezzelino's control of Verona from the south. In the east Ezzelino was opposed by the commune of Vicenza. His brother in Treviso was opposed by the da Camino (a family powerful in the country west and north of that city), the little commune of Conegliano, the anti-da Romano Trevisans, and the other small cities of the March. The most energetic threat to the da Romano's power came from Padua, which had just allied with Brescia and Mantua against Verona and the da Romano.

Da Romano control of Verona was itself of recent origin. On 14 April 1232 Ezzelino, with the support of some of the knights and people of

---

[4] On these factions, see Andrea Castagnetti, 'Appunti per una storia sociale e politica delle città della Marca veronese-trevigiana (secoli XI–XIV)', in Reinhard Elze and Gina Fasoli (ed.), *Aristocrazia cittadina e ceti popolari nel tardo medioevo in Italia e Germania* (Annali dell'Istituto Storico Italo-Germanico, 13; Bologna, 1984), 49–50. On the Vicenzan *popolo* see G. de Vergottini, 'Il "popolo" di Vicenza nella Cronaca ezzeliniana di Gerardo Maunsio', *Studi Senesi*, 48 (1939), 645–61.

[5] The words 'Guelf' and 'Ghibelline' are anachronistic for this period. In the 1230s use of these terms was restricted to Florence and Tuscany. Only in the 1240s did they come into use outside of Tuscany. Their use in Bologna begins only in the 1270s. On these terms, see Hyde, *Society and Politics in Medieval Italy*, 132–4. Furthermore, these labels would imply a clear alliance-structure which did not exist in the Veneto of the 1230s. It is better to think of our parties as groupings of local factions whose interests sometimes corresponded with those of one another, with those of the papacy, or with those of the empire. Nor was there a consistent opposition between pro-imperial magnates and pro-papal populars; in Verona, for example, the popular party was usually pro-imperial. See Castagnetti, 'Appunti per una storia', 73.

[6] Verci, iii. 110; Sutter, 95.

Verona, had captured the city's *podestà*, Guido of Rho and his judges in the Palazzo Comunale. Reinforcements for Ezzelino soon arrived under Emperor Frederick's vicar at Ostiglia and the counts of the Tyrol, Olremo, and Pianio. Ezzelino installed William of Cremona as *podestà* and crushed the remaining opposition militarily.

Those opposed to this expansion of da Romano power reacted swiftly. On 19 May the Mantuans attacked from the south but inflicted only temporary damage to some castles and bridges. Resistance also surfaced in the March of Treviso. The da Camino, having lost control of Treviso to Alberico, entered the protection of Conegliano and the bishop of Ceneda, Albert. This group then turned to Padua for protection, and took to the field against Treviso, Verona, the da Romano, and Count Guido of Vicenza.[7] A force led by Padua, which included the da Camposampiero, Azzo d'Este, the Vicenzans, and the small communes of the March, inflicted a smarting defeat on Alberico and his Trevisan supporters in late June. On 2 July Ezzelino responded to this set-back by marching with one hundred knights and one hundred bowmen to relieve Treviso, which was now gravely threatened by the rising might of the Lombards in the region. His manœuvre failed. On 27 July the combined forces of Azzo, Richard, and the da Camino clashed with Ezzelino and the Trevisans. The Lombards captured forty-eight of the da Romano knights and imprisoned them at Rovigo. Ezzelino had to retire to Verona with his remaining troops.[8] In Ezzelino's absence, the papal legates, James and Otto, had entered Verona accompanied by Count Richard of San Bonifacio. The count then made peace with the Montecchi and Quatuorviginti parties. On returning to Verona, Ezzelino immediately drove out both Richard and the legates, thereby incurring the Church's wrath and excommunication.

In October, Mantua, Padua, and Vicenza launched a three-pronged attack on the district of Verona, putting it to fire and the sword. In this attack the Mantuans received military support from Milan, Bologna, Faenza, and Brescia. Even Ezzelino's erstwhile allies, the Trevisans, took the opportunity to devastate his holdings. On 1 November, the Mantuans withdrew from the Veronese *contado* after a payment of £4,000, but it required a bloody victory by Ezzelino, Verona, and the Montecchi over the Paduans and the San Bonifacio at Opeano to force a complete withdrawal. In this battle they captured the anti-da Romano *'podestà'* in

---

[7] Alliance document published in Verci, i. 112.
[8] These developments are recorded by Parisio, 8, and in great detail by Maurisio, 26–8.

exile' of Verona. Ezzelino could now breathe more easily. Meanwhile, the Trevisans laid siege to Conegliano, failing to take it only because of the arrival of reinforcements from Richard and Azzo. The siege was lifted, with the Trevisans gravely defeated. Alberico, however, managed to rout the Paduans, capturing many of their men and imprisoning them at Bassano. Ezzelino, meanwhile, repulsed an attack by the Vicenzans.

As 1233 dawned the tide seemed to be turning.[9] Ezzelino captured and partially destroyed Richard's own castle, and Verona occupied and reduced to ashes the Lombard stronghold of Caldiero. With the struggle looking more and more even and further war more futile, perhaps the time had arrived for negotiation.

PREPARATIONS IN THE VENETO

John had left Bologna secretly after his 31 May arbitration between the commune and bishop of Bologna. His journey north took on the appearance of a triumphal procession as it approached the *contado* of Padua. His reputation as a preacher and peace-maker preceded him. At the rock of Monselice, some 15 miles south of Padua, the entire commune of that city came out to meet the preacher. They placed him in their *carroccio* and carried him into the city with great pomp.[10] The reception appears to have been prearranged; most likely the Paduans had themselves invited him to negotiate the pacification of the March of Treviso. It was a clever move, since the Paduans' military fortunes had been souring in recent months.

After a taste of John's intense peace-preaching in their own city, the Paduans handed over their disputes for arbitration.[11] This project probably occupied only a small part of his attention, since Padua had little internal factionalism.[12] Whatever local political mediation he performed, John's peace-making at Padua, and later in the March of Treviso, catapulted him into the inter-communal arena, since local factions all had connections with the larger groupings under the direction of powerful

---

[9] So Sutter, 97–8, who follows Maurisio, 30–1, and sees the entire eastern campaign as the result of Paduan plots and expansionism. The da Romano were equally expansionist. The da Romano victories of late 1232 seem tactical, not strategic.

[10] Rolandino, 45; his arrival is also briefly recorded in Maurisio, 31; and *Ann. Vet. Veron.*, 92.

[11] Rolandino, 45: '[M]ultas predicaciones fecit per Paduam.' Maurisio, 31: '[D]e omni discordia, quam habebant, compromiserunt in eundem suo arbitrio determinandis.'

[12] So Castagnetti, 'Appunti per una storia', 53.

families, such as the da Romano, the d'Este, and the da Camposampiero.[13] John made the reconciliation of the disputes between the powerful magnates of the area and the communes of the region his principal goal during the northern campaign.[14]

Within a few days, John was on his way north towards Treviso. Entering this region, John left the area of Lombard control and entered that of the da Romano. As he moved north he kept up a vigorous pace. He preached in every small town and stronghold, voicing the praises of the Virgin and her Son, the Prince of Peace. The biblical text given by the Paduan observer Rolandino to describe John's activity—'Blessed are the feet of those bringing peace'[15]—may well have been the theme of his sermons. The peace-maker gave his first solemn sermons in the March at Treviso and Feltre. There he addressed citizens of both cities, the da Romano, the da Camino, and the count of San Bonifacio, as well as citizens from Conegliano, Verona, Mantua, Brescia.[16]

John's mediation in the north resembled that at Bologna. First he received general authority to arbitrate disputes. Along with this, he received the power to revise city statutes as he wished. Although he reserved publication of his most important arbitration-decrees until a later date, he ordered the immediate release of the many prisoners taken during the last 2 years of war.[17] After preaching at Treviso he returned south, passing first to Vicenza, doubtless to prepare for similar mediation there.[18]

---

[13] On these alliances, see Gina Fasoli, 'Città e feudalità', in *Structures féodales et féodalisme dans l'occident méditerranéen IXe–XIIIe siècles* (Rome, 1986), 380–2.

[14] Rolandino, 45: '[Johannes] voluit cum auxilio Dei firmare pacem inter omnes et singulas civitates, viros potentes et nobiles Lonbardie, Marchie Romagne.'

[15] Cf. Isa. 52: 7.

[16] Maurisio, 31: '[V]enit Trevisium, fecerunt Trevisini illud idem, sic fecerunt Feltrini et Bellunenses, sic quoque domini de Camino, sic et illi de Coneglano et idem fecerunt Domini de Romano.' Godi, 10, has a more specific list: '[E]tiam Tarvisio et Feltro inter dominos de Camino, illos de Coneglano, dominos de Romano, Veronenses, Mantuanos, Brixenses et comitem Sancti Bonifacii.'

[17] Maurisio, 31–2; repeated in Godi, 10.

[18] *Ann. Vet. Veron.*, 92: '[Transtulit se] triuisium et postea vincenciam. Et post hec adiit Bononiam.' Sutter would have it that John demanded the office of 'Dux et comes' from the council of Vicenza at this time. I cannot see how this is possible. First, Maurisio, a citizen of Vicenza and an eye-witness to the events, places John's demand after Paquara. Second, John's political problems in Vicenza followed almost immediately on his assumption of power, as we shall see later. Third, the longest period John could have spent on his first trip to the Veneto would have been from 31 May to 20 June—we know he was in Bologna on those two dates. Assumption of the office would have accomplished nothing at this time, since John would have departed for Bologna immediately. Attilio Simioni, *Storia di Padua dalle origini alla fine del secolo XVIII* (Padua, 1968), 265–7, places the meeting at the Prato during this journey.

Statute revision and arbitration of local disputes would have, must have, occupied only a small fraction of John's time. During his travels north, his most important enterprise was laying the groundwork for the great assembly of the entire region at which he would publish a general peace-plan. This was to be held on 28 August at the place called Paquara near Verona. John had his plans in order for the great assembly only after the preaching-campaign to Treviso, because he did not announce the time and date of the great meeting until later, in the Prato della Valle at Padua, during a solemn sermon attended by representatives of all the major cities and powers of the Veneto.[19]

By 20 June, John was back in Bologna to revise his arbitration between the bishop and commune. His whirlwind tour of the Veneto had lasted 3 weeks at the most.[20] The stories of his recent triumphs must have preceded him, since all Bologna turned out to meet the returning hero. They placed him upon a white horse, hoisted a silk baldachin over his head on four lances, and led him into the city amid wild rejoicing.[21] After this spectacle, there are reports of the first datable resistance to the peace-maker's activities. Following his triumphal entry, John's detractors sent word to the pope that the preaching friar had usurped papal prerogatives by accepting such honours. At the papal court, William of Modena, the same bishop who had helped John escape Bologna for the Veneto, stepped forward and asked to see the pope privately. He confided with an oath on the Scriptures that John was not only a faithful servant of the Holy See but that he himself had seen angels standing at the friar's side as he delivered a sermon. The pope seems to have accepted this story, since, in Rome at least, John's reputation remained untarnished.[22]

Pope Gregory grew more and more impatient as he found himself unable to co-opt John for his own uses in Tuscany—after all, as far as the

---

[19] Maurisio, a Vicenzan, tells us vaguely (32): '[S]tatuit certum terminum pro pace facienda et firmanda apud civitatem Verone, precipiendo predictis civitatibus [Marchie] et personis quod essent tunc ibi coram eo pro predicacione audienda et laudum pacis inter omnes.' Rolandino, a Paduan, p. 45, gives us more specific information: 'Et convocatis primo principibus totius Marchie in Pratum Vallis in Padua, facta ibi predicacione sollempni, constituit et ordinavit quod in proximo futuro mense augusti covenirent in campanea Verone iuxta flumen Attatis predicti barones, omnes rectores, ambaxatores et communia civitatum.' See also Godi, 10.

[20] Sutter, 101, gives three weeks as the time expended.

[21] This incident, recorded by Th. Cant., 242, is not dated; Sutter, 102–3, suggests that it occurred at this point.

[22] Who were John's detractors? John attracted a number of adversaries at Bologna—Boncampagno of Signi and Guido Bonatti for example. This communication to the pope seems to be an attempt to scupper John's reconciliation project in the Veneto by provoking a negative papal reaction. Could this be the first example of *Paduan* hostility to John?

pope could see, there was nothing left for him to do in the north. During the first or second week of July, John received yet another letter from Gregory. This time the pope did not ask, but commanded, that he come south to negotiate peace between Siena and Florence.[23] On the same day that the pontiff dispatched that letter, 27 June, he also wrote to the bishops of north Italy, granting them the power to impose excommunication on anyone who impeded John's movements.[24] To these letters, he added another the following day, ordering in the strongest language that the *podestà* and council of Bologna give John freedom of movement. Gregory seems to have been convinced that John's lack of responsiveness to the obvious needs of the Church in Tuscany could only be ascribed to his being impeded by the selfish Bolognese. The pope's requests and letters may well have been provoked by the bishop of Modena, who, while defending John against his detractors, must also have described how he had accompanied John on his clandestine flight by night from Bologna to Ferrara. There is no doubt that Gregory trusted Bishop William. During the bishop's visit in Rome the pope relieved him of his see and promoted him to the post of legate for the Baltic Provinces.

John had now found new worlds to conquer; he simply ignored the papal request. By early July he was on the road again, this time directing his attention to those western parts of the Veneto that he had ignored during his first excursion. The destination of this journey was Verona, the centre of da Romano power.[25] Before his departure from Bologna, John or his representatives managed to convince the pope that the friar was better employed in the Veneto than in Tuscany.[26] So, on 13 July, the papal chancery authorized an indulgence of twenty days for all who in the course of a week attended three of John's incidental sermons and an equal indulgence to those who attended just one Solemn Preaching.[27] The pope had evidently abandoned his hope for a peace campaign in Tuscany.

[23] *BOP*, i. 56, letter 86 (Auvray, i. 801, letter 1436).

[24] *BOP*, i. 56, letter 85 (Auvray, i. 802, letter 1437).

[25] *Ann. Vet. Veron.*, 92: '[E]t tandem venit veronam et demum Juit Bononiam.'

[26] All reports that John preached the revival in Tuscany stem from the erroneous belief that he obeyed the papal command. Vincenzo Maria Fontana, *Monumenta Dominicana* (Rome, 1675), 31, first made this deduction; then Antoine Touron, *Histoire des hommes illustres de l'Ordre de saint Dominique* (6 vols.; Paris, 1743), i. 73. A host of later scholars, in particular Girolamo Tiraboschi, *Storia della letteratura italiana* (8 vols.; Venice, 1824), iv. 249, pick up this idea of a campaign in Tuscany. From this last it passed into Henry Charles Lea, *History of the Inquisition in the Middle Ages* (3 vols.; London, 1886), ii. 204, and finally into the *Enciclopedia italiana*, 'Alleluia'.

[27] *BOP*, i. 57, letter 88 (Auvray, i. 813–4, letter 1461).

On his journey north to Verona, John replayed his successes in Bologna and Padua on an even grander scale.[28] He preached first at Mantua, Padua's ally to Verona's south, and then began a triumphant march north. At San Bonifacio, the very stronghold of Count Richard, he met the populace of Verona, who carried him triumphantly into their city and heard him preach in the Piazza Erbe.[29] From Verona he continued his preparations to engineer a peace between the two great parties. First, he demanded oaths from Ezzelino da Romano, the Veronese *podestà* Guizardo Tealdisco, and fifteen knights each from the Montecchi and Quatuor-viginti factions. In these oaths they pledged to obey his commands and those of the Roman Church.[30] In their turn, Richard of San Bonifacio, the commune of Mantua, the da Camino, and representatives of Padua, Vicenza, and Treviso also swore to obey John's decisions.[31]

At this time, Lombard troops from the cities of Padua, Mantua, Treviso, Ferrara, and Brescia were still in the field around Verona. They still held the Veronese *carroccio*, which they had captured during their devastation of the *contado* the year before. John convinced them to return it to Verona as a good-will gesture.[32] This was John's finest hour; never again would he command such universal and unconditional respect and admiration. He had single-handedly convinced the warring factions of the Veneto and the Marches to lay down their weapons, show signs of willingness to compromise, and swear to accept his arbitration of their disputes. They took their oaths amid tears of rejoicing and praises of God.[33]

John then mounted the newly returned *carroccio* of Verona, and the citizens carried him in triumph through the city to the Piazza del Mercato ('in foro Veronae'). They called on him to name a provisional

[28] On the Verona political background, see Carlo Cipolla, *La storia politica di Verona* (Verona, 1954), 103–8.

[29] Parisio, 8: 'in foro seu mercato Veronae praedicavit'.

[30] This oath to obey the Roman see was a prerequisite for the lifting of the excommunication which Ezzelino and the citizens of Verona had incurred the previous year, when they drove out the papal legates. Gregory IX gave John the authority to lift it on 5 Aug., *BOP*, i. 59, letter 92 (Auvray, i. 823–4, letter 1488). At that time the pope gave John power to absolve any others excommunicated for harming the rights of the Church, *BOP*, i. 58–9, letter 91 (Auvray, col. 823, letter 1487). These letters would both have arrived in ample time for Paquara.

[31] *Ann. Vet. Veron.*, 92, places these oaths after Paquara. Since they are clearly the agreements to accept John as arbitrator at Paquara, Parisio, 8, Maurisio, 32–3, and Godi, 10, have placed them at the correct time.

[32] Parisio, 8: 'Et ob hanc causam Ferrarienses, Paduani, Trivisani, Vincentini, Mantuani et Brixienses post paucos dies de mandato ipsius fratris Ioannis dederunt eidem fratri Ioanni carrocium Veronensium.'

[33] Carlo Sigonio, *Historia de Rebus Bononiensibus Libri VIII* (Frankfurt, 1604), 44, says that such emotion accompanied the reconciliation that even Ezzelino was weeping.

rector for their faction-ridden city. John, sensing the desires of the crowd, proposed himself to them as *dux* and *podestà*. Shouting their approval the people acclaimed him as 'dux et rector'.[34] From this new position of authority, John set about extirpating all remaining opposition to his great project of reconciliation. He exacted oaths to accept his arbitration from all the citizens. Some did resist; whether for political or religious reasons is unclear. Whatever the case, on 21 July and the three days following John presided over the execution by fire of sixty 'heretics' who had refused to take oaths accepting him as peace-maker. Their number included men and women from the best families of the city.[35]

While John settled his affairs in Verona, his agents were active else-where putting things in order for the assembly of Paquara. Gavardo and Berardo, two Dominicans from Sant'Agostino in Padua, arranged the release of the last Trevisan prisoners still held by Conegliano, spelling out the terms in an instrument drawn up on 3 August.[36] With his power now consolidated and things proceeding smoothly, John must have en-joyed considerable satisfaction as he returned a second time to Bologna.[37] Behind in Verona he left a vicar, Vitale de Bono, to act in his place.[38]

---

[34] Parisio, 8: '[E]t super eo carrocio ipse frater Ioannis ascendit in foro Veronae, et de voluntate populi Veronensis ipse frater Ioannes se elegit in ducem et potestatem Veronae, et de voluntate populi Veronensis ipse frater Ioannes populo clamante fuit electus in ducem et rectorem Veronae.' The change in the title is not significant.

[35] Parisio, 8–9: '21. Iulii dictus frater Ioannes in tribus diebus fecit comburi et cremari in foro et glara de Verona 60 ex melioribus inter masculos et foeminas de Verona, quos ipsos condemnavit de haeretica pravitate.' Also, more briefly, in Maurisio, 33. For the identification of the 'glara' as Verona's Roman amphitheatre, see Sutter, 117. Carlo Cipolla, 'Il patarinismo a Verona nel secolo XIII', *Archivio Veneto*, 25 (1913), 64, says that the Pataria was the most prevalent heresy among the 'better families' of Verona. This sect's refusal to take oaths implies, for me at least, a political motivation for John's persecution. The exchange of the peace with heretics was also forbidden, see Elisabeth Vodola, *Excommunication in the Middle Ages* (Berkeley, Calif., 1986), 52–3. My conclusion is confirmed by the chronicle of Giacomo Pindemonte, quoted by Domenico Bortolan, *Santa Corona: Chiesa e convento dei domenicani in Vicenza, memorie storiche* (Vicenza, 1889), 101, 'Anno 1233 frater Iohannes ordinis predicatorum multos Patarinos, ac Pauperes Leoninos, in theatro comburere fecit, *quia* jurare noluerunt praecepta eius.' [Italics mine.]

[36] For some reason that evades me, Sutter, 118–20, sees this document as indicating that John moved from Verona to Padua in late July. He then arbitrarily assigns the assembly of the representatives of the Veneto in the Prato della Valle to this date. This is completely against Rolandino, 45, and *Chronicon Marchiae Trevisinae et Lombardiae*, ed. L. A. Botteghi (RIS[2] 8:3; Città di Castello, 1916), 10, both of which place the assembly in the Prato in continuity with John's preaching in the March of Treviso.

[37] *Ann. Vet. Veron.*, 92: '[Johannes] venit veronam et demum fuit Bononiam.' Although these annals place this phrase before the description of John's activities at Verona, they obviously imply that these activities occurred before his return to Bologna.

[38] As we know from documents issued by Vitale on 16 and 19 Aug. discovered in the archives of Verona and published by Sutter, 184–5.

THE ASSEMBLY AT PAQUARA

Nevertheless, in spite of his successes, all was not well. Especially at Padua, hesitation and outright resistance were increasing. At the centre of this dissatisfaction was the aristocratic Benedictine prior of San Benedetto Novello, Jordan Forzaté. With his low-key diplomatic style, Jordan was in complete contrast to the flamboyant John. Before the Dominican's arrival, Jordan had completely dominated the politics of Padua, ruling the city 'as a father would his family'.[39] Jordan was already acquainted with John, since both had served on the committee that scrutinized miracle reports during the canonization of St Anthony of Padua. Although Jordan would attend the assembly at Paquara, his later actions show that he held John's efforts in low regard. Perhaps it galled him to see his careful work to maintain the independence of Padua from the da Romano endangered by the upstart Dominican's novel version of peace. Very likely, the Benedictine was also smarting from finding his traditional authority in Veneto politics overshadowed by a mendicant demagogue. Whatever their reasons, as he and the Paduans approached Verona for the assembly of 28 August, they looked around suspiciously at the gathering crowds. They murmured to themselves that their enemies were carrying hidden arms, contrary to John's command.[40]

John's attention was elsewhere. Before leaving for Paquara, he spent some time in Bologna clearing up problems that had arisen in his absence. In particular, he wrote a report to Pope Gregory and requested faculties to absolve a certain Bolognese knight named Milancio from an excommunication incurred for damage done to the papal city of Viterbo and for defying the ban of the Church. Gregory granted him the faculties in a letter dated 25 August. In the same letter he also imposed on Milancio a punishment of 3 years' military service in the Holy Land and the payment of indemnity to the citizens of Viterbo. As the Pope drafted this letter, John himself was already back at Verona preparing for the great assembly.[41]

The site chosen for the assembly, the field called Paquara, is located

---

[39] Rolandino, 45, says of him: 'Padua . . . quam hactenus non aliter rexerat, quam regit pater familias domum suam.'

[40] Rolandino, 45.

[41] The date of John's return is unknown. It was after 19 Aug., on which date Vitale de Bono was still acting as his vicar, but before 28 Aug., the date of the assembly itself.

about 4 miles south of Verona on the banks of the Adige.[42] It lies between the small commune of San Giovanni Lupatoto and the church of San Jacopo della Tomba.[43] The field exists today, still bearing the name of Paquara.[44] Two bridges were thrown across the river to allow approaching crowds to cross.[45] Workmen constructed an enormous wooden tower-like structure, some 28 metres in height, to make the preacher visible to the crowds.[46] The date set for the great assembly was Sunday, 28 August 1233, the feast of St Augustine.[47]

As the crowds gathered, their numbers must have swollen to the tens, perhaps hundreds, of thousands.[48] In attendance were all the major ecclesiastical and civil dignitaries of the region. Among the ecclesiastics were Bertold the patriarch of Aquileia, and the bishops of Verona, Brescia, Mantua, Bologna, Modena, Reggio, Treviso, Vicenza, and Padua. Two of these bishops, Guala of Bergamo (a Dominican) and William, formally of Modena, were John's collaborators in the revival and peace-making. Each of the great ecclesiastics arrived in procession, preceded by his cross and clergy. Present also were the peace-maker's vicars from the cities of Verona, Bologna, Padua, and Parma.[49] Two of these are known by name: Bartholomew, the prior of Verona, and James Boncambio, the university professor whose spectacular conversion to the

[42] Rolandino, 45: 'Locus autem . . . distat a civitate Verone circa miliaria IIII, super Attacem.'

[43] Parisio, 9: '[I]nter Sanctum Iacobum de la Tomba et Sanctum Ioannem Lovototum super ripas Atacis in pratis quae appellantur Vigomondoni.' On the identity of Vigomondrone and Paquara, see Gino Sandri, 'Paquara e Vigomondrone', *Atti dell'Academia di Agricoltura, Scienze e Lettere di Verona*, 5th ser., 13 (1934), 101–15.

[44] See Sutter, 122.

[45] Parisio, 9.

[46] Rolandino, 45: '[Q]uadam specula, lignaminibus artificiose constructa; et hoc artificium altitudine quasi per cubitus sexaginta.' And Parisio, 9: 'Super quo belfredo ipse frater Ioannes ascendit.'

[47] The feast of St Augustine had precedence over the Sunday; see William R. Bonniwell, *The History of the Dominican Liturgy* (New York, 1944), 107.

[48] The estimate of Parisio, 9—400,000—is certainly exaggerated. But the crowd was huge. Rolandino, 45, an eye-witness, speaks of the assembly as the 'largest in the history of the Veneto'. Maurisio, 32, also an eye-witness, calls the assembly the 'largest since the time of Christ'.

[49] During his preparations John never visited Venice, but it is not impossible the Venetians sent a delegation of observers to Paquara. This should explain the reference to John's supposed captivation of the doge, which is referred to in a scurrilous song preserved by Salimbene, 78: 'Et Johannes johanniçat | et saltando choreizat. | Modo salta, modo salta, | qui celorum petis alta! | Saltat iste, saltat ille, | resaltant cohortes mille, | saltat chorus dominorum, | saltat dux Venetiarum. . . .' The scene does sound like Paquara.

religious life had enlivened one of John's sermons at Bologna. Ominously present with the Paduan clerical delegation was Jordan Forzaté.

Eyewitnesses and the documents issued on 29 August list the civil dignitaries present: Azzo d'Este, both of the brothers da Romano, the da Camino and, although not listed, certainly Richard of San Bonifacio. Present also were delegations from the communes of Brescia, Mantua, Verona, Vicenza, Padua, Treviso, Bologna, Ferrara, Modena, Reggio, Feltre, Belluno, and Parma. Among these delegations, those of Brescia, Mantua, Padua, Verona, Treviso, and Vicenza came in formation behind their *carrocci*—it would have appeared a marshalling of their troops, had they been armed. To these official delegations can be added a large number of private citizens, both men and women, who came for the spectacle and the show. All were barefoot, out of respect for the great prophet.[50]

Ceremonies preceded John's address to the crowds. They certainly included hymn-singing and, as it was Sunday, perhaps the Mass.[51] When the friar mounted the pulpit, so deep a hush fell over the crowd that all could hear his words clearly—a wonder in the opinion of the eyewitnesses.[52] John began his sermon by invoking Christ's words from John 14: 27—'Peace I leave you, my peace I give to you.' Throughout his sermon, he skilfully applied this text to himself and his project for the reconciliation of all Lombardy and the Veneto. Finally, he announced that in the authoritative judgement he was about to pronounce he also enjoyed the official authorization of the Holy See.[53]

His sermon concluded, John then promulgated his peace decrees. He opened by announcing that the parties led by Ezzelino da Romano and the Veronese on one hand, and by Richard of San Bonifacio, the Mantuans, and the Paduans on the other, were officially reconciled and were to give each other the kiss by which this reconciliation would become inviolable and perpetual.[54] His decree commanded capitulations

---

[50] On those present, see Maurisio, 32; Rolandino, 45; Parisio, 9; *Chronicon Marchiae*, 10; *Liber Regiminum Padue*, ed. Antonio Bonardi (RIS², 309); and *Ann. Brix.*, 818. The witness lists on the peace instruments of 29 Aug. published by Verci, i. 103–6, supplement these reports.

[51] Some rituals beyond the preaching—hymn-singing, for example—would have been highly likely. See Victor Turner, *The Ritual Process: Structure and Anti-Structure* (Ithaca, NY, 1982), 167–8.

[52] Maurisio, 32: 'Predicavit inter illos et—quod est mirabile—vox illius ab omnibus audiebatur mirabiliter et clarissime.'

[53] Rolandino, 45.

[54] Maurisio, 32; *Ann. Brix.*, 818; and Parisio, 9.

by both parties. First, the da Romano were to relinquish their holdings in the district of Padua that had been the cause of so much bloodshed during the previous year. In return they received an indemnity of £15,000 from the city.[55] The leaders of the two parties were then to be united by marriage: Rinaldo, the son of Azzo d'Este, was to be betrothed to Adelaide, the daughter of Alberico da Romano.[56] Since both were small children, the actual marriage would be postponed for several years.[57] He announced at least this much publicly at Paquara, eyewitnesses report. From the peace instruments themselves may be retrieved the details of John's decree.[58] Count Richard could reoccupy all of the lands seized from him during the war, as could all others who had lost land because of the conflict. Any other disputes over property John reserved to his personal arbitration.

At the end of his address the preacher declared sanctions against those who broke this peace: temporally, a fine of 1,000 gold marks; spiritually, excommunication and John's curse on their crops and lands.[59] After thundering these penalties, John turned to leave, and the crowds burst out into shouts of thanks to God. Then suddenly, as if he had forgotten something, John returned to the platform and, quieting the crowd, made one final announcement: Ezzelino da Romano was to become a citizen of Padua with freedom to enter the city.[60] As the principals exchanged the kiss, many mortal enemies sought each other out and made peace.[61] The delegations were now free to depart. They returned to their homes, still singing and shouting the praises of God and Friar John of Vicenza.

Not all were so enthusiastic and delighted.[62] Among the Paduans,

---

[55] Maurisio, 32. The *Liber Regiminum*, 310, gives the sum as £25,000. Maurisio, an eye-witness, is to be preferred.

[56] It is of note that in 1239 the Emperor Frederick II tried to arrange the same marriage-alliance in order to tie the Este to the da Romano and the imperial cause, Maurisio, 39. But again, the attempt fell through, see David Abulafia, *Frederick II: A Medieval Emperor* (London, 1988), 313.

[57] Maurisio, 32; *Liber Regiminum*, 310; Parisio, 9.

[58] *AIMA* vi, col. 641, 1171.

[59] Maurisio, 32; Rolandino, 45.

[60] Clearly not pleased with this arrangement, the Paduan Rolandino, 45, wrote: 'Ultimo, post omnia sua dicta, quasi obliviosus fuisset, addidit quod dompnus Ecelinus de Romano esset civis paduanus et reciperetur in citadanciam paduanam.'

[61] Maurisio, 32.

[62] The *Chronicon Marchiae*, 10, and the *Liber Regiminum*, 309–10, both written in Padua, sum up the pessimism about Paquara in that city. They describe John's decrees and immediately add that his power was extinguished 'cito'. The report from Padua's allied city of Bergamo, *Ann. Bergom.*, 810, and from *Chronicon Estense cum Additamentis usque ad Annum 1479*, ed. Giulio Bertani and Emilio Paolo Vicini (RIS² 15:13; Città di Castello, 1908), are

some returning home were already mouthing the complaint that their party was to be destroyed, that their enemies had attended with hidden weapons, and that they had been rushed into the peace by fear that their enemy's supporter, the Emperor Frederick, was about to return to the region.[63] The weakness of the Great Peace of Paquara was already showing itself, even as the friar proclaimed it.

JOHN'S DOWNFALL

Nevertheless, following Paquara, other parties submitted their disputes to John's arbitration. On 29 August, representatives of Padua, Treviso, Conegliano, the da Camino, and Bishop Albert of Ceneda officially selected John as 'arbiter, mediator, and friendly reconciler' of their disputes. This agreement was probably a formality, since John issued his decision on the very same day. When the Paduans heard of this decree, it must have added to their fears that John was undermining their influence and power to the benefit of their enemies and the da Romano. The decree was completely in favour of Treviso. The da Camino and the commune of Conegliano were to submit to the jurisdiction of Treviso and, although this is not explicitly stated, by so doing enter the orbit of Alberico da Romano. The dispute over episcopal lands John reserved for judgement at some future date.[64]

News of this judgement caused a sensation when it arrived at Conegliano. The outraged Coneglianese and the da Camino fired off an angry appeal to the pope, whose authority John had invoked in proclaiming his decree.[65] The prime mover of this appeal seems to have been Count Tisone of Camposampiero, an ally of Padua then holding office as *podestà* of both Conegliano and Ceneda. He called up the general assembly of Conegliano, announced the appeal, and delegated Matthew, the syndic of the city, to carry it personally to the pope. The Coneglianese drafted a second appeal on 5 September, and among its witnesses is Jordan Forzaté. Had he advised Count Tisone from the start? It is not unlikely. The defender of

equally depressing. Pietro Gerardo, *Vita et gesti di Ezzelino terzo da Romano* (Miscellanea di storia veneta, 2nd ser., 2; Venice(?), 1894), 50–1, followed by Sutter, is untrustworthy; see Girolamo Arnaldi, *Studi sui cronisti della Marca trevigiana nell'età di Ezzelino da Romano* (Rome, 1963), 212–23.

[63] Rolandino, 45.   [64] Text in Verci, i. 105–6.   [65] Text in Verci, i. 106–7.

the rights of Padua had begun his struggle to destroy John and his hated projects.

Within a few days after Paquara, John entered Vicenza. He presented himself to its council, and announced that he wanted full legal and administrative authority over the city—that traditionally attached to the offices of duke and count—to reorganize municipal affairs. The commune complied, and John received the city statute-books for revision.[66] The reasons behind John's action are clear. Vicenza was allied with Padua, and for the peace to hold the Vicenzans would have to welcome back the exiled members of the pro-da Romano factions, the Montecchi and the Quatuorviginti. These men had violently opposed Vicenza's Paduan alliance and supported Verona. For this they had been exiled and their property confiscated. The Vicenzan chronicler Maurisio explains the reason for the willingness of his compatriots at home and in exile to hand over power to John: they hoped that he would remove the ruling *podestà*, Henry de Rivoli, and install someone fully acceptable to both the pro-Padua and the pro-da Romano factions. John instead committed the fatal blunder of not naming a new *podestà* before departing for Verona.

John had left himself open to the schemes of his enemies. Taking counsel with Jordan Forzaté, the Vicenzan strongman Uguccione da Pilio and those favouring the da Camino and Padua induced Henry de Rivoli to open the city to the Paduans and fortify it in defiance of John and the Veronese. Henry's pro-Paduan government set to work drawing up the lists of those to be banned for favouring the da Romano.[67] Hearing news of these events, John dashed back to Vicenza with a small contingent of Veronese soldiers.[68] His popularity in Vicenza was still high, for outside the city a large throng of sympathetic citizens met him. Leading his soldiers and the mob, John entered the city. He began circulating through the streets, successfully demanding that the rebels hand over their weapons and fortified towers to him.

Only at one point, on reaching the house of the Zachame, did he encounter even a hint of defiance. Even this token resistance quickly melted before John and his followers. The crowd finally reached the

---

[66] For events at Vicenza in September of 1233 we are almost wholly dependent on Maurisio, 33–4. Godi, 10–11 and Parisio, 9, add little to our knowledge. Extant Vicenza statutes contain nothing identifiable as John's reforms.

[67] So Giambattista Pagliarini, *Chronica Vicentiae*, Vicenza, Biblioteca Maciana MS CCCLXVI, fo. 24, ed. and trans. G. Alcaini, *Cronaca di Vicenza* (Vicenza, 1663), 38.

[68] Parisio, 9.

Palazzo Comunale. They seized the *podestà* and his judges and put them in custody. John and the crowd fell to work destroying the hated books that listed those to be banned under the now-fallen pro-Paduan government. Against all odds, John seemed to have triumphed again by his daring and the force of his personality.

The victory was not to be. While John was still at work in the Palazzo Comunale destroying the books, those sympathetic to the fallen government threw open the city to the troops of Uguccione da Pilio and the Paduans. As quickly as it had appeared, John's following dissolved, abandoning him and his few Veronese troops. The surrounded and terrified Veronese retreated to the episcopal palace and, after a mere skirmish, surrendered. John had fallen into the hands of his enemies. Uguccione threw his Dominican captive into prison. It was 3 September, one week after the assembly at Paquara.

When news of the coup at Vicenza reached Verona, the government immediately imprisoned everyone associated with the San Bonifacio or Paduan factions. Later, after John's release and return to Verona, these prisoners were released. Release, however, came only after they had proved themselves to have been in no way implicated in John's imprisonment or the rebellion at Vicenza. Having regained their freedom, they, quite rationally, fled Verona for pro-Paduan territory.

Two days after John's imprisonment, his enemies, including Jordan Forzaté and the bishop of Padua, assembled in the piazza of the episcopal palace of Vicenza. The friar had the humiliating experience of witnessing them draft an appeal to the pope against his decrees. In it they termed his actions 'null, iniquitous, and unlawful'.[69] John remained intransigent and proudly refused to bend to their will and revoke his decisions. John sent his own message to the pope asking for help. Gregory IX did not reply until 22 September, when he sent John a letter of commiseration. The pope also wrote to the bishop of Vicenza about John's condition.[70] The letters were unnecessary: by the time they arrived, John was already free. The circumstances of his release are unknown; perhaps it was motivated

[69] Text in Verci, i. 107. The editor of Maurisio, Giovanni Soranzo, p. 33, n. 6, would have John at liberty on 5 Sept. because the events took place '*nella piazza* del vescovado di Vicenza'. I do not see what difference this makes—the drawing-up of the instrument certainly would not have been done in John's prison cell. He posits that the events leading to John's imprisonment took place only after the humiliation of 5 Sept. I think that my order of events, imprisonment then humiliation, makes more sense. It can account for Parisio's report that John was imprisoned on 3 Sept., something Soranzo must deny.

[70] Sutter, p. 136, n. 1, reviews opinions on the length of John's imprisonment.

by fears of Veronese retaliation. After returning to Verona, John finally capitulated, and on 30 September in the Dominican convent of Verona he formally rescinded his 29 August decree concerning Treviso. Completely humbled, he then had to act as witness to the instruments by which the da Camino swore fealty to the bishop of Ceneda. John's attempt to subordinate Conegliano to Treviso was utterly routed.

John still functioned as *rector* of Verona, but his power was spent.[71] On 24 September, he presided over the installation of the new joint *podestà*, Robert of Modena and Almerico of Bologna.[72] This act marked the end of his direct rule of the city. The remnants of John's peace were evaporating. Those favourable to the da Romano fled Vicenza in secret and sought cover in Verona. Rule by faction, with its consequent bannings and exiles, was returning to the region. During his last days in power John turned more and more to military force to enforce his decisions and plans. He called on Bolognese troops to garrison the imperial castle at Ostiglia and on sympathetic Vicenzans to occupy those of Count Richard and Calderia. He then successfully ordered the Vicenzans to hand them both over to the count. He soon discovered how untrustworthy his military support was. When he then ordered the Bolognese to turn over Ostiglia to Verona, they flatly refused. 'He then returned to Verona with all his intentions completely frustrated.'[73] At some time in the autumn, probably before 24 November, when Nicholas Tonisto became *podestà* of Verona, John abandoned the Veneto and returned to Bologna, perhaps visiting Brescia on the way.[74]

'And thus Uguccione da Pilio, the Paduans with the advice of brother Jordan, and the da Camino destroyed brother John's Peace, and so the deeds of brother John were considered as nothing.' So wrote Maurisio, closing his account of the Veneto peace-making campaign. Within a year the da Romano and Lombard armies fought a pitched battle on the very field of Paquara.[75] Within a few decades, brother John himself and the revival he had led disappeared into obscurity.

[71] Maurisio, 33: 'Johannes regit civitatem, sicut potest, sua potestate ibi posita pro nichillo reputata.'

[72] *Ann. Vet. Veron.*, 391.

[73] Parisio, 9: '[E]t idem reversus est Veronam totali sua intentione frustratus.'

[74] Maurisio, 33: '[D]imissis Veronensibus et Vicentinis, ivit Bononiam.' Francesco Barbarano, *Historia ecclesiastica della città, territorio e diocesi di Vicenza* (Vicenza, 1649), 89, speaks of a visit to Brescia, on which contemporary reports are silent.

[75] *Memoriale Potestatum Regiensium*, col. 1108. Cipolla, *Storia Politica di Verona*, 109.

PART II
REVIVALISM AND POLITICS

4

The Making of a Revival

EVERY revivalistic preaching-campaign of the middle ages required a preacher, but none was truly a preacher unless he had found his hearers. The medieval audience could be remarkably fickle and capricious. Some came hoping for edification, but many, possibly most, came for entertainment.[1] This impression is supported by the graphic descriptions of the French preacher of the Fourth Crusade, Foulques of Neuilly. His hearers had no reticence about expressing their expectations and demands. They brought out their sick on pallets and set them in the squares and roads through which he would pass. They strained to touch his clothing or the hem of his robe and so be healed of their ills.[2] The scene could have come from the passage in Acts which describes the preaching of the Apostles.[3]

The merely curious came because they had heard marvellous stories, hoping to see the 'prophet' perform in their own square the feats he had performed in some town down the road. The astute preacher understood his hearers' hopes and never let them go home disappointed.[4] Foulques, a consummate showman, knew the way to respond to their desires. He must have made an impressive sight. The crowds hemmed him in, pushing and shoving for a place, a chance to touch; he, always in control, immediately seized the initiative and ordered those in front to pass holy water back to those behind. Through the merits of the holy man and the unhesitating faith of the crowd many, reportedly, received healing, and

---

[1] On the theatrical in penance-preaching, see Ida Magli, *Gli uomini della penitenza: Lineamenti antropologici del medioevo italiano* (Milan(?), 1967), 96.

[2] Jacques de Vitry, *The Historia Occidentalis of Jacques de Vitry: A Critical Edition*, ed. J. F. Hinnebusch (Spicilegium Friburgense, 17; Fribourg, 1972), 97.

[3] Acts 5: 15–16: '[T]hey even carried out the sick into the streets, and laid them on beds and pallets, that as Peter came by at least his shadow might fall on some of them. The people also gathered from the towns around Jerusalem, bringing the sick and those afflicted with unclean spirits, and they were all healed.'

[4] The building of an audience through word of mouth and the importance of a preacher's reputation is clear as far back as the 11th cent. Geoffroy le Gros, *Vita Beati Bernardi {Tironiensis}* (ch. 5, §38), ed. Godefridus Henschenius (AS 11: Apr. II, 231E), reports that the hermit-preacher Bernard of Tiron's first audiences came because of his reputation as a counselor.

Foulques built his reputation. In contrast, a lack-lustre performance would spell a preacher's instant demise. A conventional or unoriginal performance brought immediate rejection. Initially receptive crowds promptly repudiated one Franciscan preacher, Claro of Florence, when he repeated the theme of a successful homily.[5]

Some took a dim view of the crowds' unending quest for a novelty and a good show. Boncompagno of Signi, the famous professor of rhetoric at the University of Bologna, disgusted by people's credulousness and their belief in the miracles reported of John of Vicenza, put it out that on a certain day he would himself take off from the hill outside of Bologna where the shrine of the Madonna di San Luca now stands and fly. The whole town reportedly turned out—old and young, women and children—for the show. After letting them stand half the day in the scorching sun, Boncompagno appeared, equipped with two paste-board wings, and announced that they could go home with the divine blessing. That the whole town should turn out for such a preposterous reason was miracle enough.[6] True or not, such a story could only circulate if it was credible that a large number of people would turn out in return for the promise of a good show.

PREACHERS AND THEIR HEARERS

Every great revivalist had a special following of devoted admirers, and possession of such a claque distinguished him from run-of-the-mill preachers. To this core following, he had to attract other hearers, until the size and anticipation of the audience reached 'critical mass'. Peter of Verona, a preacher who, like Foulques, almost always appeared in the company of a crowd, was at his best before large, enthusiastic audiences. All knew that as the number of hearers grew, God manifested ever greater

---

[5] Salimbene, 549, was not at all sympathetic to this fickle audience; it was the 'turba maledicta et simplex, que non novit legem.' (Cf. John 7: 49.)

[6] Salimbene, 77–9. In Boncompagno of Signi—see, *Testi riguardanti la vita degli studenti a Bologna nel sec. XIII (dal* Boncompagnus, *lib. I)*, ed. Vigilio Pini (Testi per esercitazioni accademiche, 6; Bologna, 1968), 37–8—there is a similar incident. There, Boncompagno reports that his form letter mocking those credulous enough to believe that magic could, among another things, make an ass fly, generated a rumour that he would perform such a feat. A crowd came to see it and stood in the sun for several hours, 'se consimiles fuisse asinis ex eo quod asinos volare credebant'. Is Salimbene's story a version of this same incident? I thank Mr Richard Guidotti for this reference.

magnitudes of grace.[7] Once in Milan the throngs were so great, and their desire to grab hold of Peter so fierce, that he had to make his way to the pulpit in a specially constructed carriage lest his devotees tear him to bits. Now that was a real audience!

Of Anthony of Padua there are similar reports. When the news spread that the famous Minorite planned to preach at Padua, crowds streamed in from neighbouring towns, castles, and villages, until the churches could no longer contain them. Anthony resorted to preaching in a wide-open field ('latissima pratorum spatia'—now known as the Prato della Valle). By daybreak, hearers who had slept in the open air and used lanterns to secure the best places already packed the meadow. They battled with late-comers to protect their seats.[8] When St Francis preached in the Piazza Maggiore of Bologna, everyone, including the members of the highest nobility, joined in the mêlée, seeking to touch him or, even better, rip off a scrap of his tunic.[9]

Foulques, Peter, Anthony, and Francis are extreme examples of the mania that a preacher could incite in his followers. Towns officially invited them to preach, and crowds long anticipated their arrival. They were among the few who could usually command an audience by the power of their names alone. Lesser revivalists strove to create for themselves this all-important reputation as a crowd-pleaser. Once they achieved this, audiences followed. Having attracted a core following, all a preacher had to do was preserve and enhance it, but this was no easy task, and one for which the revivalist might well find himself unprepared.

---

[7] Taegio (ch. 1, §9), 697E: 'Sic ergo Dominus gratiam gratiae exaequavit, quod multitudo populi ad praedicationem undique confluentis, gratiae eius magnitudinem ostendebat.' The original of this passage in the *Legenda Compilata* is Thomas Agni, *Vita Sancti Petri Martyris*, which has not been edited. The Dominicans made Agni's 13th-cent. vita the official biography of their saint. Antoine Dondaine, 'Saint-Pierre-Martyr: Études', *AFP* 23 (1953), 122–8, describes the relation of the only edited vita of Peter, that of Taegio, to Agni. Taegio compiled his *Legenda* in the 16th cent. by excerpting earlier works virtually verbatim. Dondaine, 'Saint-Pierre-Martyr', 107–9, explains Taegio's method of compilation. Since it has proved impossible to consult the manuscripts of Taegio's many sources, reference will be made to the name of the source and Dondaine's evaluation of the source's reliability.

[8] *Assidua*, 338; repeated later in Bartholomew of Pisa, *De Conformitate Vitae Beati Francisci ad Vitam Domini Jesu* (AF 4–5 2 vols.; Quaracchi, 1912–17), 268.

[9] Federico Visconti, later Archbishop of Pisa and certainly no commoner, describes with pleasure how, by God's grace, he fought through the crowd on this occasion and got close enough to touch Francis. See Federico Visconti, 'Sermones Frederici de Vicecomitibus, Archiepiscopi Pisani de S. Francesco', ed. M. Bihl, *Archivum Franciscanum Historicum*, 1 (1908), 652.

Successful preachers, those who deeply impressed their contemporaries or later merited commemoration at the altar, were not many. Some, including a number involved in the Devotion of 1233, were merely competent and merited no more than a cameo appearance in some city chronicle. The genuine stars were few. A late thirteenth-century Dominican, Jordan of Pisa, lamented, 'Today virgins are few, martyrs are few, preachers are few!'[10] This is not to say that the candidates for the honour of ushering in the New Age were lacking. Rather, they seem to have been many. But not every would-be preacher had the skill and talent to attract and keep the crowd's adoration. Popular preaching was one of the growth industries of the early thirteenth century and the mendicants were its great practitioners, but this did not mean that the new orders had developed a science of revival preaching in which they could train likely candidates. Rather, even for its practitioners, successful preaching always seemed more like an intervention of God than the result of human connivance or skill.[11]

The Dominicans—whose very name, the Order of Preachers, committed them to preaching—struggled to find suitable preachers among the many youthful candidates who entered their order. Their legislation on preaching from 1230 to 1260 reflects the problem.[12] In the beginning they had provided little in the way of training for new prospects beyond arranging for their apprenticeship as travelling companions (*socii*) of successful preachers. The primitive Dominican constitutions of 1228 left the selection of preachers to local priors for the areas under their jurisdiction, and to priors provincial for their provinces.[13] Their constitutions did envision some selection process or that, at least, some suitable (*idonei*) friars be chosen to examine candidates before they were sent out to preach. The examiners were to seek out those marked by studiousness,

---

[10] Quoted by Alexander Murray, 'Piety and Impiety in Thirteenth Century Italy', in C. J. Cuming and Derek Baker (ed.), *Popular Belief and Practice* (Studies in Church History, 8; Cambridge, 1972), 106.

[11] This attitude might be better compared to that of the American revivalists during the First Great Awakening at the end of the 18th cent. than to that of Charles G. Finney, the father of 'scientific revivalism' in the late 19th cent. See William G. McLoughlin, jr., *Modern Revivalism: Charles Grandison Finney to Billy Graham* (New York, 1959), 11–12.

[12] Heribert Christian Scheeben, 'Prediger und Generalprediger im Dominikanerorden des 13. Jahrhunderts', *AFP* 31 (1961), 140, says that later 13th-cent. legislation proved that 'Es ist nicht zu bestreiten, daß in den letzten Jahrzehnten . . . die Predigerbrüder auf der Kanzel nicht immer eine gute Figur abgegeben haben.' He places a golden age of Dominican preaching at the time of the master general Johannes Teutonicus—probably because no legislation about abuses exists for that period. I suspect it has simply been lost.

[13] On licensing preachers see Scheeben, 'Prediger und Generalprediger', 122–3.

religious observance, and the fervour of charity. Novice-masters were to instruct novices:

How they ought to be attentive in studies, whether by day or by night, in the house or on the road, whether reading something or meditating, striving to remember by heart whatever they can; and how they should be fervent in preaching at the suitable time.[14]

As for the candidates' training, it does not appear to have been very formal or elaborate. Rather, the friars appear shockingly nonchalant about it. In this they were merely following the example of their founder. Dominic, in the 1220s, sent an untrained Bolognese novice, Buonviso, off to preach. Dominic must have been a good judge of the boy's gifts, however; the novice went to Piacenza, preached, and returned with three new recruits for the order.[15]

The friars' luck, or misfortune, in attracting men like the revivalists John and Jacopino bore sometimes unappealing fruit during the generation after Dominic. Later superiors became more hesitant than the founder. Perhaps the successes of John and his followers were attracting too many inept emulators. In 1246 the Dominicans of the Roman Province had to legislate an end to the unseemly scramble for the office of preacher. They feared that the market had become flooded with cocksure young men.[16] In response, some friars ignored their superiors and ran off to solicit permission to preach from civil governments,[17] a habit that had received a boost from John's example.

Considering the preachers' lack of training, it is no surprise that the

---

[14] 'Qualiter intenti debeant esse in studio, ut de die, de nocte, in domo in itinere legant aliquid vel meditentur, et quicquid poterunt retinere cordetenus, nitantur. Qualiter ferventes in predicatione esse debeant tempore opportuno.'

[15] We have this story from the novice's own mouth. See *Act. Can. Dom.* (§19), 138–9. There is an edition of another (now lost) manuscript for St Dominic's canonization process in Jacques Quétif and Jacques Échard, *Scriptores Ordinis Praedicatorum* (Paris, 1719), i. 44–56. According to V. J. Koudelka, 'Procès de canonisation de s. Dominique', *AFP* 42 (1972), 61: 'L'édition de J. Échard . . . est à préférer au manuscrit de Venise', that is, to the edition of Walz. Since the Walz edition is far more accessible, I shall cite it when it does not differ from that of Échard.

[16] *ACPR*, 5: 'Item monemus quod fratres pro predicationibus non contendant, ne dum aliis humilitatem predicare intendimus, nostram superbiam predicemus.' It goes on to warn: 'Item monemus ne passim permittantur fratres qui non idonei predicare vel confessiones audire.' A contemporary Franciscan preacher, Luke the Lector, *Sermones de Adventu et Festivis*, Padua, Biblioteca Antoniana MS 466, fo. 198^v, testifies to the same problem when he castigates the abuses of 'vanagloriosi et pomposi predicatores qui sunt vacui et vani'.

[17] This is certainly the meaning of 'populo' in the following 1259 legislation of the Lombard Province, *ACPL*, 142: 'Item tam priores quam ceteri fratres quando debent de una terra recedere, non petant licentiam in predicatione a populo nec eis tunc se commendent.'

Dominicans had to struggle to control the abuses, scandals, and boredom provoked by unqualified, if earnest, sermonizers. With a certain resignation, the Dominicans eventually admitted that not all had an equal gift of preaching and that not all could command an audience.[18] Rulings requiring greater care in the selection of preachers appear regularly in later thirteenth-century capitular legislation.[19] Typical was the declaration of the Roman Province in 1249:

Also, we wish that for great public preachings, such as those at Rome, Florence and Pisa, as well as in other places, not any friar be commissioned but rather only famous and proven preachers from whose preaching there will be no fear of scandal, and who will be heard willingly by the people.[20]

Not all John's imitators had proved to be as skilled or effective as he. The earlier, more casual, approach had brought its own problems, and the Dominicans were now striving to remedy them.

Until well after the Great Devotion, then, the friars left the newly recruited preacher to develop his own methods and techniques. Their attitude reflected a belief that effective preaching involved a divine intervention. They placed more faith in the preachers' natural—or supernatural—gifts than in formal training.

In accord with the constitutions of the order, Dominican superiors concentrated their attention on finding those who had the 'grace of preaching'.[21] Others shared the Dominicans' belief in this grace. The Franciscan Salimbene spoke of this grace of preaching and conceived of it as a personal charisma, an extraordinary ability, unaffected by the moral quality of the preacher. He said that John of Vicenza's grace of preaching had made him so vain that he believed his miracles were an exercise of

---

[18] In 1265 the Capitular Fathers of the Lombard Province, *ACPL*, 145, wrote: 'Item predicatores non equaliter exponantur, sed illi frequentius quibus deus maiorem in predicando dedit gratiam et qui libentius audiuntur.' They repeated this legislation in 1272, when they perceived a shortage of men with a true 'gratia predicationis' to send to 'predicationes sollemnes'.

[19] A good part of this legislation concerns 'preachers general' as friars rather than as preachers since, as Scheeben, 'Prediger und Generalprediger', 127, notes, the preacher general was, 'in erster Linie Kapitular, die Generallizenz zu predigen falls sie zu Humberts Zeiten noch gegeben war, war zur Zutat.'

[20] *ACPR*, 11: 'Item volumus quod ad predicationes solemnes, sicut Rome, Florentie, Pisis, et in aliis locis, nec passim licentientur fratres nisi famosi et probati predicatores, de quorum predicationibus scandalum non timeatur et qui libenter a populo audiantur.'

[21] Dom. Con. (D. 2, cc. 17–21), 219.

personal power, not a gift of God.[22] Since grace of preaching did not imply personal holiness, it was a powerful but dangerous commodity. The grace of preaching can be defined as an extraordinary ability to sway one's hearers and an uncanny skill in sensing and controlling their hopes and fears. It carried with it the ability to inspire in hearers a belief that the preacher spoke with a special, even a divine, authority.[23] When formal training in preaching was lacking and leaders sent preachers out after a single year of theological study,[24] they could explain the appearance of truly great preachers, the 'famous and proven', only as an act of God. The crowd's enthusiasm itself testified to a preacher's supernatural election.

There was no better sign of this favour than a direct communication from God through a vision or portent. Stories reporting such events circulated for every outstanding preacher. God could point out his chosen vessel even before his birth. Dominic's mother, while pregnant with the founder of the Friars Preachers, dreamed that she had given birth to a dog that ran through the world with a torch in his mouth, setting it on fire.[25] The pious understood this as symbolic of her son's preaching and that of his order.[26] The mother of the Alleluia preacher Leo de' Valvassori received a similar portent. She dreamt that her unborn child would strengthen the hearts of Christians by feeding them with blood from the veins of his own arm. Leo, with a singular lack of humility, regularly made his mother's dream a topic of his sermons.[27]

If a preacher did not come from the womb with his divine approbation heralded by portents, signs could be delayed until the beginning of his career. Gerard of Modena was the beneficiary of one such intervention

---

[22] Salimbene, 78: 'Porro frater Iohannes de Vincentia . . . ad tantam dementiam devenerat propter honorem sibi impensum, et quia habebat gratiam predicandi, ut crederet etiam sine Deo se veraciter miracula posse facere.'

[23] As Guy Bedouelle, *Dominique ou la grâce de la parole* (Belgium, 1982), 121, notes: 'Si le prêcheur a du crédit, s'il parle avec autorité, s'il est *gratiosus*, c'est seulement par la force surnaturelle qui l'habite, s'il consent à lui donner corps.' On the phrases 'gratia praedicationis' and 'praedicator gratiosus', which first appear in biographies of the 12th-cent. hermit-preachers, see Thomas Marie Charland, '*Praedicator gratiosus*', *Revue Dominicaine* (Ottawa), 39 (1933), 88–96.

[24] Dom. Con. (D. 2, cc. 31–2), 223–4.

[25] Constantine of Orvieto, *Legenda Sancti Dominici*, ed. H. C. Sheeben (MOFPH 16: MHD 2; Rome, 1935), 288–9.

[26] The earliest version of this story is in Jordan of Saxony, *Liber de Principiis Ordinis Praedicatorum*, ed. H.-C. Sheeben (MOFPH 16: MHD 2; Rome, 1935), 27–8.

[27] Fiamma, *MF*, col. 274.

during the Alleluia itself.[28] A scoffer returned home after hearing Gerard preach in honour of St Francis and went to sleep. He then dreamt that he saw the Court of Heaven with the saints and angels venerating Christ and the Blessed Virgin. Not seeing St Francis, he sneered, 'Well, where is this Francis of whom brother Gerard spoke so highly?' An angel warned him to be silent and Christ lifted his right arm to reveal St Francis hidden in the wound in his side. Waking immediately and finding it morning, the man ran to the square where Gerard was preaching and trumpeted his vision to the crowds. Repentant, he became a Franciscan, and 'Gerard went on to work other miracles.'[29]

The truly outstanding preacher continued his mission from beyond the grave. Anthony of Padua appeared after his death to a man who had attended his sermons but remained unsure whether he should repent and confess as the preacher taught. Anthony not only warned him to repent, but told him the name of the friar to whom he was to confess. The understanding preacher, whether acting in person or in a vision, was sure to add the personal touch of directing that confession or remuneration be made to some particular individual.[30] Anthony indulged in such post-mortem spiritual direction more than once, thereby acquiring the very reputation for miracle-working that he had shunned while alive.[31]

Signs and portents could point out the possessor of the 'grace of preaching',[32] but credible portents and visions usually came only to one already known as a 'famous and proven preacher'. What was necessary to achieve this status? Salimbene's friend Barnabas was a 'proven' preacher.[33]

---

[28] The earliest version in Bartholomew of Pisa, *De Conformitate*, 263, is an admittedly late source. Luke Wadding, *Annales Minorum seu Trium Ordinum S. Francisco Institutorum* (25 vols.; Rome, 1731–1886), v. 463, dates this event to 1233.

[29] Bartholomew of Pisa, *De Conformitate*, 263: 'factus est frater minor et filius beati Francisci. Hic frater Gerardus signis claruit.'

[30] Cf. Peter Brown, 'The Rise and Function of the Holy Man in Late Antiquity', *Journal of Roman Studies*, 61 (1971), 147–8.

[31] *Assidua* (ch. 13), 346. These appearances are among the saint's oldest recorded miracles.

[32] Sansedoni, *Sermones de Tempore*, Siena, Biblioteca Comunale MS T. IV. 7, fo. 116ᵛ, alludes to this probative value of miracles when speaking of the reaction of the citizens of Nain to the miracles of Jesus: '[D]ixerunt hic est vere propheta et nuntius Dei. Unde non sunt propter hoc confessi dominum sed nuntium Dei et videtur quod bene dixerunt quia etiam multi sancti suscitaverunt mortuos . . . sicut beatus Dominicus XXXV summersos in aquam, sancta Elisabet soror regis Ungarie VII, Eliseus mortuus mortum, etc.'

[33] Salimbene, 595: 'Hic multum fuit amicus meus. Et fuit solatium clericorum, canonicorum, cardinalium et omnium prelatorum, militum et baronum et omnium qui solatium requirebant.'

He had a great facility with language, a not surprising trait in an accomplished preacher. The languages in which he preached give us an idea of where he was active. His use of Tuscan and Lombard indicate preaching in central and northern Italy, French (or most likely Occitan), in France, or probably, in this case, southern France and Piedmont. This is the same general area in which the Devotion of 1233 occurred; its preachers would have needed a similar ability. A preacher's linguistic skills expanded or limited his field of operation.

Barnabas had a special facility in speaking to children and women, an ability to win their confidence and communicate to them in a language they could understand.[34] He adapted his message to the different groups in an audience composed not only of women and children but also of rejoicing crowds of 'knights and foot-soldiers, city dwellers and country dwellers, young boys and virgins, the old and the young'. Brother Benedict played to the separate components of such an audience: he organized children into a Greek chorus that preceded him in the streets with candles and tree branches, attracting the attention of their elders, who responded with vernacular chants to his invocations and prayers, while he blew on his trumpet.[35] Such antics attracted people from far afield: 'People came from the villages to the city, under the banners of their societies, men and women, boys and girls, to hear the preaching.'[36] The ability to attract the attention of a variety of linguistic and social groups was the first step to the reputation for possessing 'the grace of preaching'.

Salimbene listed basic requirements (which he called 'marvellous') for the successful sermonizer in his description of the Franciscan preacher Berthold of Regensburg. These consisted of three hardly amazing qualities: the ability to make the sermon heard over long distances; to make those who came understand and remember it; and, afterwards, to send them home in such spirits that (at least to their own minds) they were able to complete the day's work.[37]

---

[34] '[S]cilicet qualiter pueri cum pueris pueriliter locuntur, qualiter mulieres cum mulieribus et cum commatribus suis familiari colloquio mutuo referent facta sua.'

[35] With his horn and chanting he is an archaic figure. Cf. Mircea Eliade, *Shamanism: Archaic Techniques of Ecstasy* (London, 1964), 179, on the use of musical instruments by shamans. Similar archaic elements with parallels among the shamans are John of Vicenza's control of animals, and Anthony of Padua's living in a tree.

[36] Salimbene, 70: 'Sic etiam veniebant de villis ad civitatem cum vexillis et societatibus magnis viri et mulieres, pueri et puelle, ut predicationes audirent. . . . Et cantilenas cantabant et laudes divinas milites et pedites, cives et rurales, iuvenes et virgines, senes cum junioribus.' (Cf. Psalm 148, 12.)

[37] Salimbene, 560–61.

A strong clear voice was essential, as was unusual stamina. The sermons of the revival took place in the open air and, although the chronicles no doubt exaggerate when they speak of 'crowds of hundreds of thousands, the greatest since the time of Christ', the audiences were large.[38] Beyond volume, successful preaching required a natural ability to convey meaning in spite of the defects of the medium. This meant a solid command of gesture and bodily expression.[39] It also depended on instilling in the audience a fervent desire to hear, understand, and respond. When Anthony of Padua preached, he captured his hearers' attention to such an extent that no voice or noise could be heard in the entire crowd; those present strained with all their might to catch his every word. His preaching captivated even the most venal shopkeepers, who closed their shops for the duration of his homilies.[40] There was wisdom in this. Those that bothered to remain open may well have found no customers.

CREATION OF AN AUDIENCE[41]

Before the sermon began, some groundwork was necessary. A preacher's impact depended on his audibility and visibility. Since most sermons of the Alleluia took place outside the cities, in fields or on banks of rivers, a stage of some type was essential. During the great assembly at Paquara, John of Vicenza employed a kind of look-out tower (*specula*) of wood rising almost 60 cubits—28 metres.[42] The German Bernard of Regensburg employed a similar platform, as did Gerard at Modena.[43] On

---

[38] Maurisio, 35.

[39] For example, John's turning to leave the pulpit at Paquara and then 'rushing back' to add a 'forgotten' part to his decree of arbitration.

[40] *Assidua* (ch. 13, §7), 348.

[41] On creation of mass audiences by preachers, see Ida Magli, 'Un linguaggio di massa del medioevo: L'oratoria sacra', *Rivista di sociologia*, 1 (1963), 181–98; P. di Nicola, 'Omelia come strumento di comunicazione di massa', *Sociologia*, NS, 10 (1976), 1979–97; Carlo Cipolla, 'L'omiletica nel medioevo: Teoria sociale e comunicazione di massa', *Verifiche*, 6 (1977), 298–360. The last draws heavily on Jordan of Pisa; Magli is highly speculative. Of interest because it contrasts with our preachers is Nicole Bériou, 'L'Art de convaincre dans la prédication de Randolphe d'Homblières', in *Faire croire: Modalités de la diffusion et de la réception des messages religieux du XIIᵉ au XVᵉ siècle* (Collection de l'École Française de Rome, 51; Rome, 1981), 39–65.

[42] Rolandino, 45: '[S]tetit predictus frater in eminenciori quadam specula, lignaminibus artificiose constructa; et fuit hoc artificium altitudine quasi per cubitum sexaginta.'

[43] Salimbene, 76, mentions the 'gradum ligneum', which Gerard of Modena used in his preaching but, unfortunately, does not describe it. It seems to have been portable.

another occasion, John made do with preaching from the *carroccio* of Verona. Benedict preached from the walls of a building under construction. Such devices not only raised the friar high enough to be seen, they allowed him to throw his voice over the heads of the crowd. Other physical preparations might be necessary. At Paquara, for example, workmen had to construct two bridges over the Adige so that crowds approaching from the other side could cross.[44] Although it is never directly mentioned, these preparations would have required a 'road crew' of considerable size.

The crowd at Paquara was exceptional, but even when addressing smaller groups, the friars did not leave human and material environment to chance. The simple lay-preacher Benedict may have been content with the scaffolding of a building under construction for a pulpit, but his attention-grabbing processions and chorus of children also served to set the crowds in the proper mood. As his processions wound their way through the streets, waving tree branches and carrying lighted candles, the participants were already disposing themselves to hear his message. When the preacher did not organize singing and praying before the homily, there existed other techniques for setting the audience in the proper spirit. The gathering of hearers in places somewhat distant from the city—on the banks of the Reno outside Bologna in John's case, for example—required that those listening go out of their way to attend. This created a spirit of excitement and anticipation, or even commitment, on their part.[45]

Attention to the physical and psychological surroundings of preaching did not alone create a successful campaign. The subjects of John's preachings at Bologna in 1233 reflected the economic crises and desire for peace after the many years of that city's war with Modena. It goes without saying that the preacher had to address the concerns of his hearers if he wished to be heard. But the power of the friar's preaching and his care in selecting topics for his sermons were not enough to ensure an enthusiastic response.

The preachers of the Devotion met regularly to determine the locations, days, and hours of their sermons. Then all committed themselves to the plans.[46] Such planning sometimes produced dramatic results.

---

[44] Parisio, 8.

[45] For parallels with this revivalistic preaching in 'liminal' areas such as river banks and fields, see Victor and Edith Turner, *Image and Pilgrimage in Christian Culture* (New York, 1978), esp. 1–39 and 240–1.

[46] Salimbene, 71.

Salimbene describes Gerard of Modena's most famous vision, giving us a clear picture of the kind of sacred theatre the revivalists could choreograph. Gerard was preaching in the Piazza del Commune of Parma from a wooden platform that rendered him easily visible. Then, with the eyes of all fixed upon him, Gerard suddenly stopped preaching and covered his head with his cowl. After standing as in a trance for some time, he uncovered his head and announced:

'I was in the spirit on the Lord's day', and I have heard our brother John of Vicenza preaching near Bologna on the dikes of the river Reno in the presence of large crowds, and this was the beginning of his homily: 'Blessed is the people whose God is the Lord, whom the Lord has chosen for his inheritance.'[47]

Since divinely sent revelations are not a common feature of sermons, sceptics in the audience dispatched messengers to Bologna to verify Gerard's vision. They found everything to have happened just as he had said, in fact John of Vicenza, the preacher there, had experienced a vision of Gerard himself. 'And many abandoned the world and entered the orders of the Friars Minor and of the Friars Preachers.'[48] Salimbene directly implies that the friars had concocted the vision during a planning session.[49] Indeed, the convenient occurrence of two simultaneous visions does seem too neat not to be a hoax, but its obvious fakery does not appear to have dampened popular enthusiasm for the two preachers.

Whether the preachers planned staged miracles at these meetings cannot be proven, but they certainly arranged other, less disreputable, projects. Publicity, without a doubt, headed the agenda, since a friar's reception and fame depended on spreading his reputation. Popular excitement fed by rumour, especially rumour of signs or healings, led to an increased number of excited hearers and a consequent increase in fervour. These then produced an even greater likelihood of miraculous, or supposedly miraculous, events. These led to more rumours, higher

[47] Ibid. 76–7: 'Fui in spiritu in dominica die et auscultavi dilectum fratrem nostrum Iohannem de Vincentia, qui predicat apud Bononiam in glarea fluminis Reni, et habet magnum populum coram se, et tale fuit predicationis eius initium: Beata gens, cuius est dominus Deus eius, populus, quem elegit in hereditatem sibi.' (Cf. Apoc. 1: 10 and Psalm 32, 12.) This quotation of Apoc. 1: 10, recalls the opening of Joachim of Fiore's *Expositio in Apocalypsim* (Venice, 1527), fo. 39ᵛ.

[48] Salimbene, 77: 'et secularia negotia deserentes ordinem fratrum Minorum et fratrum Predicatorum multi ingressi sunt'.

[49] The friars did not always have to concoct their marvels; an eclipse during one sermon triggered a prompt response during a preaching-campaign of 1239; see Salimbene, 164.

enthusiasm, etc. Such a mechanism came into play once the preaching and enthusiasm of the Alleluia were under way.[50] Spreading the news could not be left to chance.

At the beginning of a campaign, publicity had to start from scratch. Sometimes word of mouth was enough to get things going, sometimes not. Even Peter of Verona, proven preacher that he was, did not always find large crowds awaiting his arrival. Occasionally his first, and even his second, sermon in a city failed to bring out the hearers. Peter, then, quoting the example of the delayed response to Jonah's preaching in Nineveh, would boost the spirits of his companions and continue his sermons. Like the Ninevites, the crowds eventually turned out to hear him, at least by the third day.[51] Peter trusted in his ability to astound and excite a small crowd, and depended on the publicity provided by these few initial hearers.[52] The large company of those who followed him on the road also served as unofficial publicity agents. If the preacher provided something to recount, simple word of mouth could do much to spread his reputation. Spectators carried home tales of remarkable events during the sermon or simply described the power of the preacher's words. The stories were retold. They grew by interpretation and elaboration into great signs and marvels, and they created an environment in which the miraculous had to occur because it was an integral part of the friars' activity, just as faith-healing is part of an American camp-meeting.

The friars of the 1230s did not place their trust in word of mouth alone. The pulpit could be exploited for diffusing publicity as well as the Word of God, and the friars did this with gusto.[53] They unabashedly promoted the activities of their fellows, even those of the rival order. John of Vicenza convinced the Franciscans to include advertising for his miracles in their sermons. The chronicler Gerardo Maurisio thought John of Vicenza's reputation for marvels and the universal adoration he received rather cheaply purchased:

Well, it was no surprise; the Friars Minor were preaching publicly, as I myself heard them in the major church of the city of Vicenza, that by his intercessions

---

[50] Pierre-André Sigal, *L'Homme et le miracle dans la France médiévale XIe–XIIe siècle* (Paris, 1985), 225, notes a similar phenomenon in the case of 12th-cent. shrine-miracles.

[51] Taegio (ch. 1, §9), 697. This text is from the early and trustworthy vita by Thomas Agni. See Dondaine, 'Saint-Pierre-Martyr', 122–8.

[52] Cf. Sigal, *L'Homme et le miracle* 183–8, who identifies those healed, in particular, as the major source of publicity for shrine miracles.

[53] D'Avray, p. 4, n. 8., conceives of the preacher's control and diffusion of information as a sort of 'medieval mass media', a suggestive idea.

and prayers [John] had raised ten from the dead, and that through his intercessions the sick were being miraculously cured of their ills.[54]

Now Maurisio is not an unbiased witness—he wrote his chronicle to portray the da Romano family as the true force for order in the Veneto and John as a disturber of the peace—but his description of the Franciscan preaching rings true. John's own Dominican promoters, such as friar Thomas of Cantimpré, put out similar stories. He tells us that it was common knowledge that John had raised seven from the dead.[55] Guido Bonatti, no friend of John, tells us that the great preacher himself put it about in Bologna that the number raised was eighteen.[56] There appears to have been some confusion about the number, or, more likely, the stories in circulation grew with retelling.[57] Whatever the case, having the story in circulation was what counted.

There are similar examples of vision- and miracle-mongering wherever one looks. These often took the form of subtle (but not too subtle) propaganda for the miraculous powers of the preacher himself, or for the way in which God had especially favoured him above others. About the year 1234 one anonymous Bolognese Franciscan, cured of a bad case of diarrhoea after praying at the tomb of St Dominic, set himself to work diffusing the story of his cure and promoting the saint's cult.[58] Extant mendicant sermon-aids from the period confirm this impression. Bartholomew of Trent, a Dominican preacher and hagiographer active from the 1230s to the 1250s, compiled a collection of Marian miracle-stories for the use of preachers.[59] It brings us close to the illustrative material that the revivalists employed in their homilies. Bartholomew thought nothing of reporting on 'miracles' that he had personally experienced. Presenting himself by name, he tells how he once had visions of his dead mother while he was sick with fever, how he saved himself

---

[54] Maurisio, 32: '[N]ec mirum, quia fratres minores publice predicabant, sicut et egomet ipsos audivi predicantes apud ecclesiam maiorem civitatis Vincencie, quod eius precibus et oracionibus decem fuerant mortui suscitati et a langoribus suis infirmi per oraciones illius mirabiliter curabantur.'

[55] Th. Cant., 424.

[56] Bonatti, *De Astronomia Tractatus X*, ed. Nicolaus Prukner (Basel, 1550), fo. 210.

[57] John had some Franciscan competition in the raising of the dead. Franciscan legends say that friar Nicholas of Bologna raised the dead and gave sight to the blind in Bologna during the 1230s: *Chron. XXIV Gen.*, 225; Bernard of Bessa, *Liber de Laudibus Beati Francesci*, (AF 4; Quaracchi, 1912), 667.

[58] VF, 88.

[59] Bartholomew of Trent, *Liber Miraculorum Beate Marie Virginis*, Bologna, Biblioteca Universitaria MS 1794, fos. 70ᵛ–108ʳ. On miracles of the Virgin in Dominican tradition, see Alfonso D'Amato, *La devozione a Maria nell'Ordine Domenicano* (Bologna, 1984).

from drowning by invoking the Virgin, how an angelic guardian protected him from robbers, and so on. Bartholomew was not a Solemn
Preacher, nor one of the great, but his willingness to put himself forward
as a beneficiary of divine intervention gives credence to the jabs of the
critics who say that Leo de' Valvassori preached about his mother's pre-
natal vision or accuse John of Vicenza of promoting his own personality
cult.

There were other ways to create sensations which did not require
obvious self-promotion. Some of the revivalists had an uncanny ability to
pull off the conversion of notable figures during their homilies. This
made for great theatre.

The famous master Roland of Cremona, while at the height of his fame
as a teacher at the University of Bologna, chose the end of a homily by
the Dominican Reginald of Orleans to present himself and declare his
intention to become a friar. Reginald took off his own tunic on the spot
and clothed the postulant, an *ad hoc* schola of assisting Dominicans
intoned the 'Veni Creator', and a sacristan ran to ring the church bells.
'A great number of men and women came running to see what was
happening, and the whole city was thrown into commotion.'[60] In a
similar fashion, the Bolognese master, Moneta of Cremona, converted
to the religious life during another of Reginald's sermons. Moneta had
even been putting it out to his students that he would never go to
one of Reginald's sermons for fear of being seduced into becoming a
Dominican.[61] The crowds listening to Reginald were especially large on
the day of Moneta's conversion, since rumour had leaked out that he
planned, at last, to attend. During the Alleluia another Bolognese
master, James Boncambio, arrived at a sermon by John of Vicenza riding
a white horse and clothed in all the finery of his station and then, just as
John finished speaking, declared his conversion to the religious life. With
the crowds in tow, he rode the short distance from the Piazza Maggiore
to the friars' Church of San Nicola delle Vigne and received the Dominican
habit.[62]

These conversions seem too beautifully orchestrated to have been
unplanned and spontaneous. When the hearers who witnessed them left
the piazza, they certainly had plenty to report to those who had stayed

---

[60] *VF*, 26–27: '[F]it concursus ingens virorum ac mulierum et scolarium et tocius
commocio civitatis.'

[61] *VF*, 169–70.

[62] Borselli, *Cron.*, 22; we find other examples of conversions during preaching, for
example, during that of Jordan of Saxony at Vercelli in 1229–31, *VF*, 174–5.

home. If the preacher provided his hearers with stories to tell, half of his publicity needs were fulfilled.

Who made up the crowds attracted with so much care and effort by the preachers of the revival? This is difficult to determine with precision. Most reports stress quantity, not quality. Non-Italian chroniclers, like Matthew Paris,[63] and homilists, like Anthony of Padua, suggest that north Italian church-attendance in general, and sermon-attendance in particular, were very low.[64] The hearers' motivations differed and the extent to which they internalized what they heard varied, but more seems to have been retained than one might suspect.[65] The jabs of foreigners and the castigations of homilists are not to be taken too seriously. Observers of the Alleluia tell us that a preacher was heard 'by all classes, ages, and sexes', but this means there were many people present—it is not a sociological analysis. There is good reason to believe that sermons, or those of star preachers at least, were well attended, and attended by people from diverse states of life.

Even nobles and wealthy merchants would turn out to hear a celebrated preacher.[66] The high-born Federico Visconti came to Bologna specifically to hear and touch St Francis. An irreligious magnate like the vicious Ezzelino da Romano took care to show up for at least those of John of Vicenza's sermons that touched on the politics of his region. A Bolognese university professor of wealthy family, the already mentioned James Boncambio, not only attended John's homilies but chose one such occasion for his conversion to the religious life.[67] An extraordinary preacher attracted the lower nobility and *valvassori* as well. One castellan who had, without knowing his identity, seized and imprisoned Berthold of Regensburg, on learning who his prisoner was, immediately ordered him to deliver a homily.[68] Does this mean that wealthy Italians displayed equal interest and regularly attended the sermons of the Alleluia preachers? Probably not, but when the thrust of the preaching touched

---

[63] Matthew Paris, *Chronica Maiora*, ed. Henry Richard Luard (Rolls Series 57, 7 vols.; London, 1872–3), v. 170, where Italian city-dwellers are defined as at best semi-Christian, voluntary outsiders to Christianity.

[64] So Murray, 'Piety and Impiety', 94.

[65] Colin Morris, *The Discovery of the Individual, 1050–1200* (London, 1972), 73, sees a rather high level of internalization. Ambrose Sansedoni, fo. 39ʳ, seems to have expected, or at least hoped, that his hearers would both understand his homily and put it into practice.

[66] Contrary to Murray, 'Piety and Impiety', 93.

[67] Borselli, *Cron.*, 22.

[68] Salimbene, 560.

peace-making and politics, as it often did in early thirteenth-century Italy, self-interest dictated their attendance.

The presence of the rich and powerful did not exclude that of the poor.[69] Observers tell us that these came *en masse* to the sermons of 1233. One impoverished woman even gave birth during a homily by Jacopino of Reggio. The preacher responded accordingly. He pleaded her needs eloquently. And the pile of donations collected from the crowd—shirts, dresses, etc.—grew so large that she needed a donkey to carry it home. Those who brought out their sick to Foulques or to Anthony of Padua included many who were not well-to-do; preaching in the squares and on street corners was calculated to attract the attention of the unemployed and beggars, especially if they had heard reports of an earlier crowd's unexpected generosity in response to the pleas of the preacher.

Among hearers of such diverse class and station, who composed the core of the audience during the Great Devotion? There are firm grounds for speculation. Speaking of John of Vicenza, the *Cronaca Rampona* of Bologna tells us: 'All the residents of the city, *contado* and district of Bologna, and in particular the members of the Società delle Armi of Bologna, so believed in him that they followed him, his preaching, and commandments, with their crosses and banners.'[70] The Società delle Armi consisted of those tradesmen and smaller merchants who made up the peace-keeping forces of the city.[71] These corporations had won a role in the communal government during the uprising of the popolo in 1228. Socially, they represented the middle and upper-middle strata of the Bolognese population.[72] They would have especially responded to the preachers' message of unity and peace: anyone who might ensure unity within their communities and remove the danger of new factions was sure to receive their careful attention. The friars' appeal was not restricted to

---

[69] Brenda Bolton, 'Innocent III's Treatment of the Humiliati', in C. J. Cuming and Derek Baker (ed.), *Popular Belief and Practice* (Studies in Church History, 8; Cambridge, 1972), 79, considers them the mendicant preachers' major devotees.

[70] Rampona, 102.

[71] Gina Fasoli, 'Le compagnie delle armi a Bologna', *L'Archiginnasio*, 28 (1983), 163–4.

[72] Bonatti, *De Astronomia Tractatus*, fo. 211, describes John of Vicenza as pandering to the disreputable rabble, by which he means the Società delle Armi. These groups were *not* badly off. This charge of demagoguery was an old one against itinerant preachers and need not be taken too seriously. Already in the 11th cent., Bishop Marbod of Rennes, *Marbodi Redonensis Episcopi Epistolae*, PL 171:1484A, (Epistola 6), chastised the hermit-preacher Robert of Arbrissel for preaching on the defects of the clergy to the 'vulgares turbas'. Such phrases often mean not the down-and-out but minor artisans and those not self-employed. Carlo Delcorno, *Giordano da Pisa e l'antica predicazione volgare* (Florence, 1974), 66, believes urban artisan and mercantile classes were most attracted to mendicant preaching.

residents of the city alone. Their audience was not entirely urban, if such a word is meaningful in an age when the larger part of any city's population consisted of transplanted countryfolk.[73] As noted earlier, throngs came from the surrounding *contado* and the rural villages, often in procession with the banners of their associations.

The friar's followers were a disturbingly promiscuous mix of age and rank. When the chroniclers claim that both young and old attended the friars' preaching, they describe the reality. Benedict Cornetto, the Emperor Frederick complained, enlisted young children and other irresponsible persons to participate in processions or singing, thus disposing them against the imperial government. Even more disturbing to the critics was the unsegregated presence of women. The friars had already achieved notoriety for their appeal to women. It became, even for some of the friars' followers, a source of consternation and anxiety.[74] The Dominican *Vitae Fratrum* tells of a pious woman who doubted whether young Dominican preachers (wearing such a comely habit!) could keep themselves pure in the midst of routine contact with women. A nocturnal visit from the Virgin Mary with a promise to protect the young preachers' chastity finally silenced her reproaches.[75]

Preaching to women generated other, even more serious, resistance. At least in common report, impious and jealous husbands regularly prevented their wives from attending sermons.[76] Peter of Verona, who attracted hosts of female admirers, once healed a woman mutilated by her husband for attempting to escape and hear his homily.[77] The preacher's

---

[73] Lester K. Little, 'Evangelical Poverty, the New Money Economy and Violence', in David Flood (ed.), *Poverty in the Middle Ages*, (Werl, 1975), 13.

[74] A complaint going back to the 11th cent.: Geoffroy of Vendôme, 'Epistola Fratri Roberto' (Epistola 47), PL 157:182A, explicitly criticized the hermit preacher Robert Arbrissel's dangerous familiarity with women. And in the 12th cent., Bernard of Clairvaux, 'Sermo LXV', in *Sermones super Cantica Canticorum*, ii, ed. J. Leclercq *et al.* (Sancti Bernardi Opera, 2; Rome, 1958), 174–6, attacked the unshaven and unwashed clergy who abandoned their churches to wander about in the company of women. They were, he said, 'the little foxes destroying the Lord's vineyard' of the Song of Solomon.

[75] *VF*, 40–1.

[76] A story told of, among others, John of Vicenza (see Th. Cant., 424) and no less than three times of Anthony of Padua (*Chron. XXIV Gen.*, 137 and 138–9, and Bartholomew of Pisa, *De Conformitate*, 267). In these stories the wife always miraculously hears the homily over a long distance, allowing something to spill or be ruined as she does. The husband is then converted by the intervention of a miracle. Either he too hears the homily over a great distance or the spilled wine or oil is miraculously restored. The literary origin for such wine-restoration miracles may be Gregory the Great, *Dialogues* (1. 9. 4), ed. Adalbert de Vogüé (Paris, 1979), ii, 78–80.

[77] Taegio (ch. 4, §28), 702E, original from the miracle collection known as the *Miracula Collecta de Mandato Berengarii Magistri Ordinis*. This collection was compiled at the order of

female adherents could react in ways that scandalized their husbands. When Peter of Verona arrived at Cesena, not only the usual vulgar masses, but even 'noble and very honourable matrons' (*nobiles et honorabiles valde matronae*) joined the mob running to see him. They came in such unseemly haste that they neglected to put on their over-mantles (*chlamydes*), a garment without which no honest woman of the region dared appear in public. Peter had not even planned to give a sermon that day, but the mob dragged him to the principal square of the city, set him in a prominent place, and forced him to preach the 'Word of the Lord'.[78] Solid and respectable citizens may well have wondered what crazed mania had taken hold of their womenfolk.

For all their success at attracting hearers, the friars of 1233 encountered some pockets of resistance. Noticeably absent in the descriptions of their audience is one of the major components of the pre-popular ruling class: bankers. Absent also, and perhaps overlapping with bankers, are usurers, in particular those publicly recognized as such (*usurii publici*). These groups had good reason to shun the preachers, who railed against usury and greed.[79] Also seemingly absent were lawyers and judges, members of a profession that predominated in the old urban leadership. Like usurers, they had good reason to absent themselves. John of Vicenza was patronized by the very associations of the *popolo* which had just broken the aristocratic jurists' traditional monopoly on government. It must have galled the legal profession to watch rank amateurs usurp their traditional roles as arbitrators and mediators.

When a revivalist found hostility in the cities where he preached, it came principally from those who had traditionally dominated communal politics. Resistance arose in different sectors of the population in each particular city. In Padua, for example, the commune as a whole received John of Vicenza with enthusiasm; it was a religious, Jordan Forzaté, the

---

the Dominican chapter at London, 1314, and dates to approximately 1316. On this source, see Dondaine, 'Saint-Pierre-Martyr', 128–30. Dondaine, 146–7, considers this collection of miracles the nearest thing to a contemporary narrative source for Peter Martyr.

[78] Taegio (ch. 2, §16–18), 699A–B, originally from Pietro Calo, *De Sancto Petro Martyris* from his *Magno Legendario*; see Dondaine, 'Saint-Pierre-Martyr', 130–2. Calo's legend is among the latest of Taegio's sources, dating between 1323 and 1340. Dondaine, 137–9, considers this story relatively reliable, unlike Calo's other stories about Peter at Cesena which he places 'au rang des fables'.

[79] John of Vicenza, for one, made them a special target of his tirades, as implied by Borselli, *Cron.*, 22. Ambrose Sansedoni died of a burst blood-vessel in his neck while fulminating against usurers.

Benedictine prior, who before John's arrival had enjoyed the most influential voice in the communal government, who orchestrated the resistance and eventually succeeded in bringing about John's downfall. In Emilia and the Romagna, John's two most vocal critics were an astrologer, Guido Bonatti, and a rhetoric professor, Boncampagno of Signi. Does this suggest that the intellectuals of the period rejected and opposed the demagoguery and emotionalism of the revival?[80] Probably not. The learned could be caught up in the excitement along with the illiterate. John of Vicenza numbered a Bologna professor among his conversions, and the Dominican master general of the 1230s, Jordan of Saxony, had great success among the Bolognese university population, bringing into the Dominican order, among others, Master Roland of Cremona.

It was not the friars' vulgarity that offended their critics. Guido bitterly complained that the powerful sought advice from friar John, a mere rabble-rouser, rather than from himself, a real expert in astrology and politics. His animosity is manifestly personal. Boncampagno seems to have been offended by credulousness and enthusiasm in themselves. Whatever motives Guido and Boncampagno had for opposing John, their opposition does not imply a general rejection of the preachers by the intellectuals.

Attracting and pleasing an audience was only the first step to success. Once hearers had gathered, the preacher's words had to bind them into a unity—a group that sensed its identity and was willing to put the preacher's programme into action. There were different ways of doing this. Creation of this cohesion began with the sermon itself. Ambrose Sansedoni once proposed that in modern times holiness had declined in the Church. At least in comparison with saints in the ancient days of the martyrs, the moderns cut a poor figure. Ambrose provided an explanation:

We see that the more one abounds in temporal goods, the less one abounds in spiritual ones. In the case of the primitive and modern Church, [we see] that when, a long time ago, she was poor in temporal goods, she was rich in spiritual ones. So, all the popes up to Saint Silvester were saints and all the bishops were of nearly equal quality.[81]

---

[80] As Fumagalli, 261, suggests.

[81] Sansedoni, *Sermones de Tempore* fo. 135ʳ: 'Videmus qui plus in temporalibus habundat minus habundet in spiritualibus. In ecclesia primitiva et moderna, quia quando diu fuit paupera in bonis temporalibus tunc erat dives in spiritualibus. Unde omnes pape usque ad sanctum Silvestrum fuerunt sancti et omnes quasi equipollentes episcopi.'

The notes for this sermon contain no specifics on the state of the modern clergy and hierarchy—perhaps Ambrose passed over them in discreet silence—but if a modern reader can imagine his implication, so could his thirteenth-century audience.

Prelates do not generally cut a good figure in Ambrose's preaching. In his notes for the parable of 'The Tares Sown among the Wheat' he wrote:

Or, the sleep [of the owner of the field] can be understood as the negligence of prelates, who, if they were not asleep, would be engaged in the correction of vices and the punishment of evil-doers. . . . So, if the popes and cardinals of the last two hundred years had bothered to attack their causes, this land would not be so filled and polluted with wrongful gains and other sins.[82]

It is hard to tell how these notes translated into a homily, but no doubt hearers went away with the feeling that rich negligent prelates were responsible for a good number of the world's ills. There is no quicker way to unity than to focus hearers' attention on a villain or common enemy.

Preaching against powerful malefactors verged on demagoguery, and contemporaries labelled it as such. For the hearers, on the other hand, these condemnations must have been satisfying news to hear, and the temptation to pander to their prejudices must have been great. In 1240 the Dominican general chapter found itself forced to take measures to curb the abuse. The capitular fathers admonished preachers 'that they diligently warn people to honour churches and prelates so that they observe the rights of local churches.'[83] This general directive does not seem to have had great effect, since in 1246 they issued another, more forceful decree:

Let the brethren be careful that in their preaching they not seem to be speaking of any particular person. . . . Any who are offenders in this are to be suspended by their priors from the office of preacher and given a suitable penance for such time as the gravity of their excess requires.[84]

---

[82] Ibid., fo. 28ᵛ: 'Vel sompnum potest intelligi negligentia prelatorum, qui, si non dormissent, in correptione vitiorum et persecutione malorum non [*sic*] essent. Mala et peccata in proposito. Unde si papa et cardinales, qui fuerunt iam duocento anni, voluissent obstare principio, ista terra non esset ita malis lucris plena et contaminata et sic de aliis peccatis.' The confused grammar is typical of Ambrose's sermon notes.

[83] *ACG*, 15: 'Fratres nostri diligenter admoneant populum. ut ecclesias et prelatos honorent. ut ecclesiis reddant iura sua.'

[84] *ACG*, 36: 'Caveant fratres diligentissme. ne in predicacionibus aliquam personam notabiliter tangere videatur. . . . Qui vero reprehensibiles fuerint; a suis prioribus cum condigna penitencia, ab his officiis ad tempus suspendantur. prout eorum maior. vel minor requirit excessus.'

This prohibition did not completely prevent the friars from attacking unpopular figures. Church authorities themselves pointed out some enemies—usurers and heretics, for example. Frederick II found himself the victim of hostile preaching by one preacher of the Devotion. He complained to the pope that Leo de' Valvassori's homilies were unjustly inciting the people of Milan against him.[85] In the wake of Frederick's long estrangement from the papacy and his recent excommunication, one might wonder how promptly and effectively Gregory responded to the appeal.

Reviling the sins of the powerful, the wealthy, and the outsider is too constant a theme in revivalistic preaching to be ignored, but it was not its essence.[86] The preachers were not mere rabble-rousers or mob-agitators. Nor was violence against usurers and heretics their goal. This is evident from a careful reading of the reports that have come down to us. The vices of the audience itself were just as often the chief object of the preacher's fulminations. The violent denunciation of sin and evil could even be forged into a remarkable community-forming tool. Foulques of Neuilly shows us how this could be done. He often flew into such fits of rage while preaching that he would break out in curses and call down the wrath of God on sinners. He singled out particular recalcitrants in the audience as the special objects of his wrath; when he cursed them, they fell to the ground, writhing and foaming at the mouth, claiming that they were possessed by demons. No one was safe from his attacks. If someone disturbed his sermon, even by quiet private conversation, he would suddenly turn on him and deliver a series of curses. The crowds collapsed in terror, and there was immediately 'magnum silentium'.[87] This kind of excess, Foulques's generally violent temper, and his flights of rage brought him some criticism, but everyone had to admit his success in reforming usurers and prostitutes and in inciting men to take the cross. He collected apprentice preachers, whom he taught to preach in the same style.[88]

The jaws of hell were ever present, ready to devour sinners. A vivid experience of hell-fire is an excellent motive for repentance. Bartholomew

---

[85] *BOFM*, i. 324 (g).

[86] On attacking deviants as a source of unity in times of social unrest, see Little, 'Evangelical Poverty', 12.

[87] Jacques de Vitry, *Historia Occidentalis*, 98.

[88] Alberic of Trois Fontaines, *Chronica*, ed. P. Scheffer-Boichorst (MGH.Ss. 23; Hanover, 1904), 876–7.

of Trent, in his collection of sermon illustrations, tells us that he knew a man who had, when ill, a vision of the fires of hell and the souls of the damned writhing in agony, tormented by the demons. On recovery, Bartholomew tells us, the man 'did penance and, by my counsel, joined the Order of Preachers and therein served God and his Mother until his death'.[89] The same emphasis on divine retribution and punishment runs through several of Bartholomew's other exempla.[90] Nor is this a particularly Dominican trait; the prayers of Anthony of Padua reportedly provoked a dream of hell in a runaway novice who had stolen a valuable psalter. The youth returned, terrified and contrite.[91] The preachers rendered the reality of hell and damnation more vivid by recounting revelations and visions. In one of his sermons Leo de' Valvassori described how he had been called to hear the confession of 'a very powerful man of Milan with a great reputation for sanctity'. The man promised to return after his death and report his final fate. The man died and Leo, along with two other friars, began an all-night vigil to wait for the man's return. After telling his confreres to wake him should something occur, Leo dozed off into a light sleep. Suddenly a wind blew up and a fiery globe descended from the sky on to the roof of the cell. Awakened, Leo heard moaning and groaning. Questioned by Leo, the voice reported that he had been damned for allowing two abandoned children to die unbaptized and never confessing the fault. Leo, horrified by the sin, told the spirit to depart and 'to go his own way'. The spirit did so with shrieks of pain.[92]

A collective experience of horror, a sense of the proximity of damnation—made more vivid by the preacher's cursing or by the howling of the obsessed—typifies the friars' preaching. In his famous description of Berthold of Regensburg, Salimbene reports that when that friar spoke of the Last Judgment and damnation the crowds 'shook like a reed in the water' and that they would 'beg him for the love of God not to speak on these things, for they were overcome with fear and terror when they heard him'.[93] Their terror, however, did not stop them from coming back to

---

[89] Bartholomew of Trent, *Liber Miraculorum*, fo. 72ᵛ (miracle 140): 'Penitentiam egit et de meo consilio predicatorum ordinem intravit et ibidem usque ad mortem Deo et sue Matri perfectum servivit.' This theme of a brush with hell as the cause of repentance is common in Bartholomew: for example, fo. 73ʳ (miracle 145) and fo. 73ᵛ (miracle 151).

[90] e.g. ibid., fol. 97ᵛ (miracle 191), or fo. 81ʳ⁻ᵛ (miracles 193 and 194).

[91] *Chron. XXIV Gen.*, 133.

[92] Salimbene, 74.

[93] Salimbene, 560. D. L. d'Avray, *The Preaching of the Friars: Sermons Diffused from Paris before 1300* (Oxford, 1985), 61, writes, with reference to this passage of Salimbene, 'One is

hear more.[94] This sort of preaching retained its popularity well after the 1230s, with similar effects on its hearers. At the sermons they came perilously close to condemnation, but then, by acts of conversion, they were ritually cleansed. The social unity found in the identification and excoriation of absent evil-doers was secondary to the unity that resulted from the collective confession of sin and its rejection.

SOURCES OF THE REVIVALISTS' AUTHORITY

Not all the fear inspired by the revivalists was spiritual in origin. Sometimes it arose from the preachers' peculiar relation to physical power and force. Barnabas successfully cultivated connections with the wealthy and powerful. This was a way to find handy friends and protectors, but it was one which the friars' superiors frowned on and forbade.[95] More commonly, as with John and Ambrose, the force they enlisted was that of the hearers and townsfolk themselves. John was not a man to be trifled with. He could move an audience in frightening ways. We remember how, after one especially stirring homily against usury, the enraged Bolognese populace stormed to the house of a notorious public usurer, Pasquale Landolfo, and, after helping themselves to his possessions, burned the house to the ground—destroying his records in the process. They would have killed him, too, had he not escaped.[96]

---

liable to get an over-dramatized picture [of the preaching of the friars], based on descriptions of exceptional preachers like Berthold von Regensburg.' To d'Avray, narrative descriptions of the preachers are at best deceptive. I am not so sure. Berthold is indeed exceptional, but exceptional in the sense that he was a practitioner of a very great ability, one others would have wanted to copy. Similar dramatic elements are present in descriptions of much less famous preachers.

[94] Dare I compare this sort of preaching to a horror movie? The viewers cover their eyes and recoil in terror at the very parts of the movie they came to see. McLoughlin, *Modern Revivalism*, 510, considers the creation of anxiety and 'psychological tension' essential to the success of revivalism.

[95] Some Dominican friars spent too much time frequenting the courts of the powerful. In 1240 the Dominican general chapter issued the following decree, *ACG*, 15: 'Fratres curias regum vel principum absque magna necessitate. seu fructu animarum. frequentare non presumant.'

[96] Borselli, *Cron.*, 33. The link between John's preaching and this riot has been debated because only one author makes the connection, the Dominican Borselli. Furthermore, three of the reports of the riot, Rampona, 101; Matteo Griffoni, *Memoriale Historicum de Rebus Bononiensium*, ed. Lodovico Frati and Albano Sorelli (RIS[2] 8:2; Città di Castello, 1902), 9; and Villola, 101, place it at the end of 1232. Borselli, *Cron.*, 23, and Bolognetti,

John allowed his hearers to respond to his message in symbolic ways, such as by communal acts of penance and by triumphal processions.[97] He had an uncanny ability to turn these public acts of piety into displays of power. John always appears with a crowd, and this crowd is not harmless. At Bologna he was accompanied by the Società delle Armi, and they attended him with banners waving. The society's statutes rigorously controlled the use of their banners. It restricted their use to periods when they were called to arms to defend their quarters or to quell riots. The very carrying of a banner in public on other occasions was strictly forbidden, since it would be taken as a provocation or a call to arms.[98] His critic Guido Bonatti says that John's followers protected John with pikes and mercilessly beat anyone who tried to approach the preacher. Guido claims, no doubt with exaggeration, that some were killed, many were wounded, and John took great pleasure in the fray.[99] In short, John travelled about Bologna accompanied by a mob in military array.[100]

Paradoxically, as a preacher's fame grew, so did the dangers to, and limits on, his authority. Miracle stories could take on a life of their own, and the preacher risked getting swallowed up in the image he had created. Since this image was one of miraculous power he risked becoming a mobile, almost impersonal, point of contact with the Divine. The preacher could become so much the thaumaturge that his essential activity of preaching could be forgotten. Mobs chased and manhandled Foulques of Neuilly, trying to rip off bits of his clothing as relics.[101] More than once, he laid about with his walking-staff to beat back the

---

Cronaca detta dei Bolognetti, in CCB, ed. Albano Sorbelli, (RIS² 18:1:2; Città di Castello, 1911), 102, place it at the end of their description of John's activity in the year 1233. I would trust Borselli since he is a trustworthy source, and if the riot occurred in winter (the most likely time since the first three chronicles connect it with food riots), the confusion over whether the year began on 1 Jan. or 25 Mar. may underlie the discrepancy in the date.

[97] On extra-liturgical rites as expressing popular fears and hopes, see Raoul Manselli, *La Religion populaire au moyen âge: Problèmes de méthode et d'histoire* (Montreal, 1975), 113–14.

[98] The statutes of the Società delle Traverse di Porta S. Procolo imposed a fine of 10s. bol. for displaying the banner in the streets when there was no call to arms. They also imposed a heavy fine on the standard-bearer who left the banner locked up and unavailable while out of town, so important was it for the collecting and rallying of the troops. See Bol. Armi Stat., 140.

[99] Bonatti, *De Astronomia Tractatus*, 182; on this passage, see Sutter, 88–9.

[100] Similar orchestrations of force are found for other 13th-cent. preachers, e.g. Ambrose Sansedoni, in Alessandrino (ch. 6, §42–6), 190A–F.

[101] We find such hunting for the relics of preachers as early as the 12th cent.: crowds pulled hairs from the crusade preacher Peter the Hermit's mule, see Guibert of Nogent, *Gesta Dei per Francos*, PL 156:704. In 1233 people collected pieces of John of Vicenza's tunic and locks of his hair after he renewed his tonsure, Salimbene, 78.

crowds. They reportedly took the blows without murmur, counting and kissing the bruises as relics, and came on for a second try at his clothing. On one such occasion the exasperated Foulques, showing both his sense of humour and his command of the situation, announced, 'Do not tear my clothing, it is not blessed!' He then blessed the cape of one of the most avid relic seekers. The crowd then fell on the poor man and ripped his clothes to shreds.[102] Peter of Verona attracted an equally persistent favour-seeker at Cesena. As he crossed the square to preach, a young nobleman, John di Biagio by name, accosted him. The youth had suffered for some 10 years with a tumour on his hand and he begged a cure. The preacher, obviously exasperated, declared that a cure was not within his competence and flatly refused to perform one. The boy persisted, so Peter, with no little disgust it seems, blessed the tumour and told the boy to have faith in God. He then marched off to begin his sermon. Poor Peter! The boy was miraculously healed and soon arrived to interrupt the sermon with shouts of thanksgiving.[103] Miracle-working might do wonders for reputation; it did not always assure audiences attentive to the message of the sermon. The prudent preacher kept the hope for miracles and wonders within bounds. It is not surprising that Anthony of Padua rebuffed requests for cures.

The preacher's persuasive power itself presented a temptation to the preacher. His ability to control audiences, to diffuse a reputation, and to draw and unite a crowd could blind him to the reality that he was effective only as long as he gave voice to the unexpressed needs of his hearers. The preacher could not move hearers against their will, because a preacher's power over his audience arose from his ability to sense and respond to the hearers' deeper needs and hopes. When Salimbene accused John of Vicenza of arrogance and the belief that he could work miracles without divine grace, he was not far wrong. John's downfall came because of his failure to understand the limits and origin of his own power. The citizens of Vicenza had invited John to their city, 'hoping that he would remove the _podestà_ and replace him with another acceptable to all, by whom both factions of the city would be governed'. They found, however, that once they had given him power he did no such thing.[104] Now, the removal of a _podestà_ is no mean feat, but John could accomplish

---

[102] Jacques de Vitry, _Historia Occidentalis_, 98.

[103] Taegio (ch. 1, §13), 698, original from the more-secure part of Calo's report on Peter at Cesena. See Dondaine, 'Saint-Pierre-Martyr', 137.

[104] Maurisio, 32.

it because he represented the wishes of the people. When he chose to ignore their desires, the people rioted, and he ended up in gaol. Any preacher who forgot that the power that he had so carefully crafted depended on the consent of his hearers did so at his own peril.[105]

---

[105] Roland of Cremona, the second preacher at Piacenza in 1233, made just such a miscalculation, see *Ann. Plac. Guelf.*, 668.

5
The Revivalist as Miracle-Worker

SALIMBENE tells us that the preachers of 1233 'put themselves forward as working miracles'. He even gave it as the distinguishing mark of that revival while reminiscing about the later founding of a lay confraternity, the Militia of Jesus Christ, at Parma:

Also I remember that an order was founded at Parma at the time of the Alleluia, that is, at the time of the other Great Devotion, the one when the Alleluia was chanted and the friars and preachers put themselves forward as working miracles, in the year of our Lord 1233, at the time of Pope Gregory IX.[1]

That a certain miraculous quality surrounds revivalistic preachers is not surprising, nor is the report that people occasionally credit them with healings or seemingly miraculous conversions. Salimbene implies more than that. His report, which other independent witnesses confirm, is that the preachers consciously and deliberately exploited their reputation for working miracles and that they made miracle-working an integral part of their preaching. This exploitation of the miraculous is remarkable, even shocking when contrasted with the long-established tradition of the holy man as a reluctant thaumaturge.[2] This new emphasis on the miraculous in preaching was not restricted to the revival of 1233. By mid-century a distinguished Dominican preacher could couple miracle-working with preaching and persecution as the marks of a true apostle.[3] In the light of

---

[1] Salimbene, 467: 'Item recordor, quod ordo iste factus fuit in Parma tempore Alleluie, id est tempore alterius devotionis magne, quando cantabatur "Alleluia", et intromittebant se fratres Minores et Predicatores de miraculis faciendis, anno Domini MᵒCCᵒXXXᵒIIIᵒ, tempore pape Gregorii noni.'

[2] See Benedicta Ward, *Miracles and the Medieval Mind: Theory, Record and Event 1000– 1215* (Philadelphia, 1982), 175. His fellow monks rebuked St Bernard for pride when he worked miracles, or so his biographer claims: William of Saint-Thierry, *Sancti Bernardi Abbatis Clarae-Vallensis Vita et Res Gestae*, PL 185: 253, 315.

[3] Ambrose Sansedoni, *Sermones de Tempore*, Biblioteca Comunale MS T. IV. 7, fo. 89ᵛ. For the link between miracles and preaching in the life of Christ, see ibid., fo. 12ᵛ. This union was not uncontroversial. William of Saint-Amour, a contemporary of the Alleluia preachers, wrote sceptically (echoing Rom. 10: 15), *De Periculis Novissimi Temporis*, in Max Bierbaum (ed.), *Bettelorden und Weltgeistlichkeit* (Münster, 1920), 13: 'Pseudopraedicatores, qui praedicant non missi, quantumcumque literati et sancti sint, etiamsi facerent signa vel

the emphasis laid on it by contemporaries, the role of miracles in the preaching of the friars and the popular image of the revivalists is of central interest to this study.

Even scoffers and enemies of the revival recognized the importance of miracles during the Alleluia. A song current in the sceptical circles surrounding the Emperor Frederick II parodied the activities of two leading figures of the Alleluia, John of Vicenza and Jacopino of Parma. It lampooned the friars' miracles: their claims to have raised the dead, healed the sick, cast out demons, cleansed lepers, and freed prisoners from confinement. The author gave particular attention to their claims to have changed water into wine, to have spoken with angels, and to have predicted future events with precision. In fact, the waggish author wrote, he wanted to catalogue all their miracles, but his pen gave out after the sixth verse.[4] Although this ditty was unabashed parody, it reflected accurately the popular image of the Alleluia preacher: except for the reference to the 'water made wine', other, sympathetic, writers recount every other miracle held up to ridicule.

THE TESTIMONY OF MIRACLES

During the investigations into the sanctity of Dominic conducted for his canonization in 1233—4, one witness, Stephen of Spain, then prior of the Dominican house in Bologna, testified that he believed the successes of

miracula, missi autem non sunt nisi qui ab ecclesia recte eliguntur.' On William, see Tadeusz Manteuffel, *Naissance d'une hérésie: Les Adeptes de la pauvreté volontaire au moyen âge* (Civilisation et Sociétés, 6; Paris, 1970), 86. In response to such objectors, Sansedoni could, of course, quote Mark 16: 17—18, where the followers of Christ are marked by their miraculous powers.

[4] *Poésies populaires latines du moyen âge*, ed. Édélstand du Méric (Paris, 1847), 170. The song, sometimes ascribed to Pier della Vigne, reads: 'Sunt ab eis mortui plures suscitati, | Caeci, surdi debiles, infirmi sanati, | Fugatique demones, leprosi mundati, | Et aperti carceres, naute liberati. Et omnes audivimus aquam factam vinum | per Iohannem scilicet et per Iacobinum, | quod gustatum fuerat per architriclinum; | sic fecisse legimus beatum Martinum. | Loquebatur Dominus eis, cum volebant, | et ad eos angeli boni descendebant, | et mali similiter eis apparebant, | qui suis per omnia mandatis favebant. | His nunquam apostoli fecerunt maiora, | sed nec his similia, nam quacumque hora | invocabant Dominum fratres, sine mora | fiebant miracula laude digniora. | Visiones aliquas per raptum viderunt, | sed non licet homini loqui que fuerunt; | de futuris etiam plura predixerunt, | que, sicut predixerant, ita contingerunt. | Signa quidem plurima sunt ab eis facta, | que fuissent omnia hoc scripto redacta; | sed, cum vellem scribere, penna fuit fracta, | et bissexti numerus crevit in epacta.'

the Alleluia—the revision of statutes according to the moral reform preached by the friars, the return to the Church of more than a hundred thousand people who were tending toward heresy, the reconciliations of feuds, and the establishment of peace—to be the result of Dominic's intercession. 'Asked why he believed this, he responded that [these things had happened] after the time when John of Vicenza had begun to preach about the divine revelation that he had received about brother Dominic and after he had begun to announce his life, activities, and holiness to the people.'[5] The remarkable aspect of this testimony is not the miracles worked by Dominic, but the routine way in which Stephen takes John of Vicenza's revelation for granted. John, to his enemies and his friends, was unquestionably a man marked out by his special powers.

John's reputation spread far beyond Lombardy.[6] Not long after his death in the 1260s, Ambrose Sansedoni of Siena could say in one of his homilies that John was popularly known as 'of Bologna', not because he was born there but 'because of the many miracles he worked there'.[7] During his lifetime John even merited a letter from Pope Gregory IX congratulating him on the marvels ('prodigia') which God had worked through his hands.[8] The thirteenth-century sceptic and enemy of John, Guido Bonatti, reported that he had tested and examined the results of that preacher's purported miracles. All, he concluded, were faked or based on false reports. Guido was not the only sceptic who wanted to check the validity of the friars' miracles.[9] Whatever modern observers, under the influence of scepticism and rationalism, might think of miracle-working, it surely characterized the preachers in common perception. In short, whether they believed in the friars' miracles or not, all

---

[5] *Act. Can. Dom.* (§39), 158; Jacques Quétif and Jacques Échard, *Scriptores Ordinis Praedicatorum* (2 vols.; Paris, 1719), 54: 'Interrogatus quare hoc credit? Respondit quod [facta sunt] ab illo tempore postquam f. Joannes Vicentinus coepit predicare revelationem sibi de f. Dominico divinitus factam, et vitam et conversationem et ipsius sanctitatem populo nunciare.'

[6] For example, Matthew Paris, *Chronica Maiora*, ed. Henry Richard Luard (Rolls Series, 57; 7 vols.; London, 1872–83) v, 146; Alberic of Trois-Fontaines, *Chronica*, ed. P. Scheffer-Boichorst (MGH.Ss. 23; Hanover, 1904), 933.

[7] Sansedoni, *Sermones de Tempore*, fo. 133ʳ: '[S]ed propter multa miracula que ibi faciebat'. This parallels a turn of phrase in Salimbene, 595, used of another Alleluia preacher: 'Frater Iacobinus de Parma, qui in Regio miracula faciebat et ideo de Regio dicebatur, et de ordine fratrum Predicatorum erat.'

[8] *BOP*, i. 55, letter 88 (Auvray, i. 713, letter 1270).

[9] For example, sceptics tested the miraculous powers of Theobald of Albinga, *VF*, 225; of Isnard of Chiampo, *VF*, 227; and (in an admittedly late and borrowed passage) of Peter of Verona, Taegio (ch. 2, §18), 699C–D (taken from a probably fabulous section of Pietro Calo according to Dondaine, 'Saint-Pierre-Martyr: Études', *AFP* 23 (1953), 138–9).

contemporary observers focused on them as the distinguishing mark of an Alleluia preacher.

What is reported explicitly for John of Vicenza—that he was viewed as a miracle worker—may, indeed must, be inferred for the rest of the revivalists. The kinds of miracle-stories repeated about the preachers are of great importance for evaluating the social and religious significance of the power ascribed to them. Some caution is in order, however, before conclusions are drawn about the hopes and fears of those who told these stories. One occasionally senses that much of the popular interest in their miracles flowed less from piety than from a curiosity to see what tricks the next round of preaching would entail or who would get the better of whom. The reports do not always or necessarily reflect a strong desire to touch the sacred; nor do they necessarily reveal some occult aspect of the preacher's charisma. They indicate, as often as not, a simple delight in the marvellous and remarkable.

John of Vicenza had a friend with a pet, a talking magpie, and the great preacher was quite taken with this bird. One day—the year was 1231—John arrived to visit his friend, only to find that the magpie had disappeared. John called out, 'Where are you, my little friend? Where are you?' To which the bird replied from the stomach of a servant, 'I'm in here!' The mortified servant fell down and begged John's forgiveness for having cooked and eaten the magpie for lunch. Word of this spread throughout the city, and for many days afterward the crowds came to hear the bird talking to John from the servant's stomach.[10]

How seriously can the modern historian take a story like this? Since it is repeated by John's admirer Thomas of Cantimpré, one might be inclined to believe that some actual event—perhaps a display of ventriloquism?—lies behind the story. Had it been a fabrication of John's enemies, intended to mock him, his followers would never have repeated it. Whatever the case, it does suggest the kind of miracle that the friars' followers delighted in. The popularity of the preachers and their miraculous performances did not always spring from deep spiritual needs or hopes.[11] There was a strong element of the theatrical, a pleasure in

[10] Th. Cant., 424.
[11] Ida Magli, *Gli Uomini della penitenza: Lineamenti antropologici del medioevo italiano* (Milan(?), 1967), 125, taking these miracle stories a bit too seriously I think, sees in them a profound crisis, an unconscious doubting of Christianity, and ultimately a testing of God. We might equally explain them as a survival of a literary type of trivial or comic miracle, the *joca*. These serve as literary relief in hagiography; they do not have serious theological import. On this genus of miracle-story, see Ward, *Miracles*, 211.

amusing and entertaining the audience, and a degree of playfulness, in the marvels of these preachers. Likewise, one can never be sure what original incident lies behind each story. In fact, a story told may have been borrowed from those of another miracle-working preacher. That a miracle may be a hagiographic stereotype or modelled on a biblical story does not necessarily strip it of significance. The origin of the story, be it some event in the preacher's life or a literary source, is less important than its content and the fact that observers were willing to repeat it about a preacher. Even more important are the circumstances in which the friars' miracles were believed to occur and what the reports themselves can reveal about the popular image of the preachers and the needs, hopes, and fears of those who followed them.[12]

The revivalists' miracles as a whole present some interesting patterns, and their mode of transmission is itself very suggestive. The early thirteenth century saw a rise in the number of miracle-collections for 'living saints', projects compiled with a view to future canonization.[13] The revivalists of 1233 themselves indulged in systematic miracle-collecting.[14] But there is no sign that anyone ever systematically collected the miracle-stories of the individual revivalists themselves, at least not until well after their deaths. The stories initially circulated individually and were preserved in chronicles and other, commonly secular, sources. This is significant, since those who transmitted the stories did not intend to prove the sanctity of their miracle-worker to ecclesiastical authorities; rather, they reported a story because they found it interesting and meaningful for them. Unlike hagiographic miracles, the miracles reported do not prove the 'Christ-likeness' or personal holiness of the preacher. As a whole, what they proclaim above all is that the preacher in his act of preaching was an instrument of God and a source of power.

No less suggestive than their mode of transmission, are the types of miracle-stories reported. Table 1 displays a typological analysis of the

[12] Howard Clark Kee, *Miracle in the Early Christian World: A Study in the Socio-Historical Method* (New Haven, 1983), 2, writes, 'It is by no means adequate for the historian to ask merely whether miracles can or cannot occur. What is reported to have taken place can never be separated from the larger framework of meaning, the assumptions about reality, the values, the attitudes toward evil, the hopes of deliverance that comprise the network of assumptions in which the experience took place and in which the event is recounted.'

[13] Ward, *Miracles*, 166.

[14] For example, Leo de' Valvassori collected testimony about the miraculous conversion of a gambler who had knifed an image of the Blessed Virgin. See Paolo Sevesi, 'Beato Leone dei Valvassori da Perego dell'Ordine dei Frati Minori, arcivescovo di Milano (1190?–1257)', *Studi francescani*, 2nd ser., 14 (1928), 47. Bartholomew of Trent was a professional miracle-collector.

TABLE I. *Comparison of Miracles by Type*

Type of Miracle	11th–12th cent.[a] (%)	Alleluia (%)
Healing and resurrection	244 (32.9)	5 (13.5)
Obtaining children	12 (1.6)	0 (0)
Protection from danger	38 (5.1)	3 (8.1)
Freeing of prisoners	5 (0.6)	0 (0)
Favourable intervention	104 (14)	1 (2.6)
Glorification of saint	27 (3.6)	8 (21.6)
Chastisements	48 (6.4)	7 (18.9)
Visions	187 (25.2)	9 (24.3)
Prophecy, clairvoyance	76 (10.2)	4 (10.8)
TOTAL	741	37

[a] The statistics for the 11th and 12th centuries are from Pierre-André Sigal, *L'Homme et le miracle dans la France médiévale (xie–xiie siècle)*, (Paris, 1985), 290–1.

miracle-stories linked to preachers of the Great Devotion. With one exception, these figures are for miracle-stories told about those friars whom Salimbene specifically named as preachers of the Devotion. In order to provide a larger sample, they also include those of Peter of Verona. He, although not named by Salimbene, was a contemporary with a marked similarity to the Alleluia preachers.[15] In any case, the only significant proportional change that results from including him is an increase in the number of healings (from 1 to 5 examples). This tabulation counts only those miracles that are transmitted as the subject of a miracle-story. There are, for example, reports that read like this one for Isnard of Chiampo:

There was in the convent of Pavia a Brother Isnard, a religious man, both fervent and a graced preacher, through whom God worked many miracles that are confirmed by witnesses: among these, five cripples recovered use of their limbs, four deaf their hearing, two mutes their speech, and three blind their sight by the touch of his hand and his invoking of the name of Jesus.[16]

[15] I have, however, omitted miracles in Taegio's legend which Dondaine, 'Saint-Pierre-Martyr', 134–61, considers very late or dubious.

[16] VF, 227: 'Fuit in conventu papiensi frater Isnardus vir religiosus et fervens et graciosus admodum predicator, per quem Deus multa miracula fecit per testes fideles probata, in que claudi V gressum, surdi IV auditum, muti II loquelam, ceci III visum, III manus usum ad tactum eius et invocacionem nominis Ihesu Christi plene recuperaverunt.' On Isnard and his miracles, see Rodolfo Maiocchi, *Il beato Isnardo da Vicenza O.P. e il suo apostolato in Pavia nel secolo XIII* (Pavia, 1910).

Such reports are significant because they show us that a miraculous aura surrounded the revivalist. They tell us little about how spectators viewed the thaumaturge and what his popular reputation was. Since the intent of this analysis is to discover what kind of miracles came to mind when contemporaries spoke of the miracle-working preachers, the tabulation includes only those miracles that exist in narrative form. This follows the practice of Pierre-André Sigal, whose figures for eleventh- and twelfth-century miracles are presented for comparison.[17] In his tabulation Sigal counted only miracles that gave some information about the persons affected and the circumstances; he ignored simple listings like that for Isnard. He also divided ante- from post-mortem miracles. Since all the stories for our revivalists report miracles worked while the friars were alive, his figures for ante-mortem miracles are alone employed.

The friars' miracles fall into nine categories. Sigal's division of life-miracle stories and post-mortem miracles allows us to compare live friars' miracle-working with what was 'typical' in stories told of living holy men in the earlier period. Of particular interest in the table are the changes in four of the categories employed.[18] There are four categories of miracles in which the difference is statistically significant. Two of these categories show a marked decline in numbers. The most striking characteristic of this comparison of miracle-stories is the lack of healing-miracles ascribed to the preachers. This lack is remarkable because most medieval miracle-stories, like those told in this century, describe miraculous healings. Healing is the miracle *par excellence*.[19] The percentage of such miracles is

[17] Pierre-André Sigal, *L'Homme et le miracle dans la France médiévale (xie–xiie siècle)* (Paris, 1985), 290–1.

[18] Standard deviations for these changes follow Joseph L. Gastwirth, *Statistical Reasoning in Law and Public Policy* (San Diego, 1988), 213–15: for standard deviations

$$Z = \frac{p_1 - p_2}{\sqrt{(p_c[1 - p_c][1/n_1 + 1/n_2])}}$$

where Z = standard deviations; p = the percentage of that type of miracle in the set of all miracle-stories considered, divided by 100 (e.g. 20% = 0.2); p_1 = this statistic for that type of miracle among earlier miracle stories; p_2 = this statistic for that type of miracle among Alleluia miracle stories; p_c = this statistic for that type among all miracle stories considered; n_1 = total number of earlier miracles under consideration (741); n_2 = total number of Alleluia miracles under consideration (37).

According to the Z test, if standard deviation is greater than 1.96 or less than −1.96, then the difference is considered statistically significant at the 95% confidence level. The results for the four categories of miracles under consideration are, thus, statistically significant at the 95% confidence level or above.

[19] As noted by Sigal, *L'Homme et le miracle*, 227: 'Le miracle type du Moyen Âge est, comme de nos jours, le miracle de guérison, bien que de façon moins exclusive.'

remarkably low, less than half that of the earlier period. If our statistics are accurate, they would suggest a variance of 2.47 standard deviations from what appears normal for the eleventh and twelfth centuries. Even a lesser decline than this would suggest that in the eyes of contemporaries the miracle-type most common in any period, healings, was in no way that most identified with the miracle-working preachers. Likewise, there is a drop in favourable interventions: a variance of 1.98 standard deviations from the norm. Since such interventions are usually gratuitous gifts for individual suppliants, it can be inferred that the friars were not thought of as easily available sources of supernatural help for private individuals. Their interest to the observer must have been on the societal, not the personal, level.

The drop in examples of these normally common miracle-types is compensated for by the increase in the incidence of two types of miracle. The first increase is in the number of miracles that serve no function other than glorifying the holy man. For the preachers of the Alleluia, this figure is 5.14 standard deviations above the norm. Such miracles reflect on the preacher alone; they do not directly benefit anyone else. Along with visions, this sort of miracle accounts for nearly half the stories reported. This leads to the conclusion that the popular image of the preacher was especially that of a powerful divine agent, privy to divine guidance. Furthermore, God acted directly to point out this quality. Another compensating change is the considerable rise in the proportion of chastisement miracle-stories. The figure for the Alleluia preachers is here 2.91 standard deviations above the norm. Such miracles indicate that God not only points out his agent, he protects him (sometimes in dreadful ways) from scoffers and enemies. There is no significant change in the percentages for giving of children, protection from danger, freeing of prisoners, visions, or prophecies. It is notable, however, that these last two categories remain large.

Such a constellation of miracle-stories as that found for the revivalists of 1233 is in accord with what Howard Clark Kee's studies of early Christian miracles would lead us to expect for an audience whose major concerns were unstable conditions in society and crises in the economic and political order.[20] In his study of ancient Christian miracle-stories, Kee divided miracles into seven groups according to the needs of those who experienced or reported them. Nearly all the miracles of the friars

[20] See Kee, *Miracle in the Early Christian World*, 293–6.

fall into his seventh category, that indicating a crisis of authority or structure in political or ecclesiastical society. This category of miracle includes those that validate the saint as a leader, portents or signs that point out the holy man, displays of personal power, and punitive miracles. Particularly absent among the friars' miracles are those that indicate a desire for spiritual healing (e.g. those that minister to the personal needs of the witness) or those that indicate desire for union with God (such as those giving a personal benefit with cosmic dimensions, e.g. exorcisms).

Following Sigal and Kee, this statistical comparison suggests that the miracles recounted indicate a perception of the preachers as loci of special power for the remedy of certain societal ills—lack of political legitimacy, chronic violence, and confusion of the social order. Beyond such general themes, it also suggests that a closer examination of the friars' miracles might provide a morphology of the authority they were perceived to possess and how they exercised it. The analysis that follows considers the more common types of miracle-stories in turn and relates them to these general themes and the revivalists' particular activities. Since the most common stories are those that glorify the holy man's power or show him as being privy to divine enlightenment, these two areas will be considered first and will provide a key for understanding the meaning of the friars' other miracles. The next most common type of story, one that reveals a more dangerous and threatening side of the revivalists' persona, one in which someone is punished for ignoring or mocking him, will then be considered.[21] Taken as a whole, these narratives not only provide a vivid and consistent reconstruction of the popular perception of the revivalists of 1233; they suggest what their audiences expected from the revivalists whom they revered.

THE CHARISMA OF THE PREACHER

Other than visions, the most common miracle-type recorded for the friars of the Alleluia is that in which the preacher displays extraordinary power over natural events or forces. These miracles are even more significant in that nearly all are performed 'on stage' while the friar was preaching to

[21] For an anthropological study of such miracles, see Kee, *Miracle in the Early Christian World*.

the crowds. Such displays set the preacher and his preaching in a cosmic arena and marked him as a special agent of God.

Peter of Verona was once preaching at Florence in the square called the New Market (now the Piazza della Repubblica) before a large and, reportedly, devoutly attentive audience. It happened that, on the edge of the crowd, there appeared the servant ('puer') of a Florentine nobleman. The boy rode a great war-horse. Suddenly—under the influence of the Devil—the horse threw its rider and bolted towards the crowd. The crowd panicked and, because of the great number present, could not make way for the horse. The preacher, spreading his arms and loudly invoking the blessing of God, ordered the crowd to sit and promised divine protection against the raging animal. The biographer tells us that the horse ran over the heads of the crowd, striking the bodies of the faithful with the full force of its hooves. Having passed from one side of the piazza to the other, it disappeared as quickly as it had come. The terror passed; no one was harmed.[22]

We are surely safe in assuming that an actual interruption of a sermon by a runaway horse gave rise to this story. The report that the interruption of the sermon was the work of the Devil reflects the crowd's perception that the Evil One was prone to disrupt preachers' holy work by his power over natural forces.[23] A preacher was pitted against him in a contest with cosmic dimensions. A revivalist may have preached against vice, usury, heresy, or urban violence, but behind those evils lurked the forces of darkness. God could and did intervene to protect his servants. Gratian, the Franciscan who admitted Anthony of Padua to the order, was preaching one day in the Romagna before a large crowd when it suddenly began to thunder and lighten. Such a wind blew up that even the men present were about to flee in panic. The preacher called in a loud voice to those fleeing, 'Do not flee, my brothers, the Lord will provide weather suitable for the preaching of his Word!' Then, with all marvelling ('mirantibus omnibus'), the oncoming rain divided and passed around the crowd so that, for a stone's throw around, the sky remained

---

[22] Taegio (ch. 3, §20), 700A–C. This miracle is followed by another, almost identical, in which a horrible pitch-black horse attacks a crowd attending a sermon, this time in the Old Forum at Florence. Here Peter's blessing causes the horse, which is in fact the Devil, to disappear. Taegio takes both these stories from Pietro Calo. Dondaine, 'Saint-Pierre-Martyr', 148–50, considers the first story ancient and the second a doublet. This story immediately calls to mind the similar scene in the recent film *Gandhi*.

[23] e.g. *Chron. XXIV Gen.*, 10–11, where the Devil wants to interrupt the homilies of Anthony of Padua.

clear, with the sun shining. 'What had occurred they ascribed to divine power.'[24]

This aura of power over nature might extend beyond the preacher to those who used his name. A certain rustic admirer of John of Vicenza was ploughing his field when he spied a magnificent eagle resting on the ground. 'Stay, O eagle, stay, I abjure you by brother John, and let me capture you,' said the rustic, and he picked up the eagle without difficulty and brought it to John who was preaching nearby. The eagle became John's mascot, following him from town to town, listening to his preaching. Then, when John had given the final blessing, it rose majestically in flight, 'in his own way giving thanks to his Creator' for the sermon.[25]

In these stories one glimpses the stuff out of which the preacher built his reputation for miraculous power. All of these incidents—omitting, of course, the story of the horse stamping on the bodies of the crowd without harm—could be explained naturally. What the three show above all is the preacher's remarkable control over his audience. In the first two, one sees the hearers' trust in his word and instruction, and their willingness to situate his actions in a cosmic struggle against infernal forces. In the third, one sees their belief that they had come into contact with an awesome yet readily available source of power. All highlight the context in which the miracles occurred: the giving of a sermon. The miracles are part of the great drama of preaching; as Salimbene said, the audience saw the preachers, and the latter presented themselves even as they preached, as miracle-workers. The two activities blend into an indivisible unity.

When one considers the preachers' power over nature, it is well to remember that popular recognition of this charisma was not spontaneous. The audiences' perception was itself the work of the preachers. A masterful preacher could make miracles happen when he needed them, that is, at the time of his sermon. A great preacher could turn occasional events—bad weather, wild horses, etc.—into a demonstration of his authority by a display of courage and insight. This skill reveals itself in action in a touching incident told of Anthony of Padua. Anthony's preaching was interrupted by the raging and shouting of a 'stultus', almost certainly someone mentally disturbed. Anthony called the man

---

[24] Thomas of Pavia, *Dialogus de Gestis Sanctorum Fratrum Minorum*, ed. Ferdinand M. Delorme, (Bibliotheca Franciscana Ascetica Medii Aevi, 5; Quaracchi, 1923), 94: 'divinae virtuti adscriberent quod fiebat'.

[25] Th. Cant., 424.

forward and presented him with his own Franciscan cord. The man quickly became docile and quiet.[26] Was he touched by the saint's gentleness and compassion, or had a demon been exorcised? The observer, considering the supernatural struggle that was the context of preaching, would have to declare for the latter.

Miracles that occurred during preaching, then, are the sort that Salimbene had in mind when he said that 'the preachers presented themselves as working miracles'. Alleluia cures, when found described, are a part of preaching; they were not private favours to the pious. Their context gave them a special added dimension, that of validating the preacher's words. Cures were the miracles that the devout and the curious turned out to see, bringing with them their sick, but, in contrast to 'nature miracles', cures were not lavishly described. There are brief résumés. The Franciscan William of Cordella was preaching in the main square of Tuscanella, when a blind boy rushed up and begged him to bless his eyes. He did, and the boy recovered his sight. The same day a man who had been bent over with back trouble for 6 years approached the preacher. William blessed him and he recovered.[27] The healings seem almost too routine and uninteresting for the chronicler to bother recording in detail.

When the reporter gives details, these often deflect attention away from the miracle itself to some other aspect of the story. Among the rare descriptions of a healing that gives concrete details is one told of Saint Peter Martyr.[28] While he was preaching in the Dominican church of Sant'Eustachio in Milan, the people brought him a mute boy and begged a cure. Peter stuck his fingers in the boy's mouth and demanded, 'What do you have in your mouth?' 'Your fingers', the boy replied. This healing is above all a response to the demands of his audience. Other healings like it occur more at the insistence of the crowd than at the preacher's connivance. As with the preacher's protection of crowds from storms and

---

[26] *Chron. XXIV Gen.*, 129.

[27] Thomas of Pavia, *Dialogus*, 250–1. Gregory IX charged William, his trusted penitentiary, with peace negotiations in Tuscany and France. Peter of Verona healed cripples too, see Taegio (ch. 4, §34), 704C–D, originally from Thomas Agni; see Dondaine, 'Saint-Pierre-Martyr', 134. Anthony of Padua also healed cripples, at least in the later hagiography, *Chron. XXIV Gen.* (ch. 23–4), 137. Healing was part of a preacher's 'job description'.

[28] Taegio (ch. 1, §11), 698, originally from both Thomas Agni and Pietro Calo. Dondaine, 'Saint-Pierre-Martyr', 147–8, suspects that this story is the older doublet of the healing of a mute child, Taegio (ch. 3, §28), originally from *Miracula Collecta de Mandato Berengarii Magistri Ordinis*.

wild horses, a judicious prayer or laying-on of hands at the demand of the people could, under the right circumstances, establish or confirm a great reputation. The preacher had to know when to consent to the crowd's requests and when to refuse.

Had the preacher remained simply a point of contact with divine protective or healing power, he would have differed little from a miracle-working shrine or the relics of a saint—of interest to people but of little direct influence on social and political policy. The Alleluia friars transformed their reputation for miracle-working into a charisma more useful in a social reformer than healing—that of prophecy. This development shows itself most clearly in the attitudes expressed towards John of Vicenza. No mere miracle-worker, John was a prophet: the 'Great Prophet of Bologna'. 'Prophet' is the most common title granted the preachers of the Alleluia. For some authors the word 'prophet' alone was sufficient to explain and describe them.

When they gave him this title, John's contemporaries did not intend the kind of prophecy that scholars associate with the name of the friars' near-contemporary, Joachim of Fiore; that is, an inspired interpretation of the Bible, giving predictions about the future or the ages of the Church. That practice is singularly lacking in stories of the preaching friars. In the entire body of texts dealing with itinerant preaching examined, one finds this variety of prophecy only once, and that well over a century before the Devotion of 1233. It is in the biography of the German hermit-preacher, Norbert of Xanten. Here, not the preacher but two canons claimed to have a prophetic key to the books of Daniel and the Apocalypse. Norbert sharply rebuked them and proved that their knowledge came from commerce with diabolical powers.[29] Nor did the friars' contemporaries intend by the word 'prophet' what the Dominican Hugh of Saint-Cher—who by timely accident or design produced a tract on prophecy at Bologna in the year 1233—meant: one who foretells the future and does so accurately.[30]

In contrast to these understandings of prophecy, Luke the Lector, a Franciscan preacher at Padua in the mid-thirteenth century, gives us a more cogent definition of the word 'prophet'. Prophets, he says, are those who 'contemplate the hidden mysteries of God and, inasmuch as they see

---

[29] *Vita Norberti B*, ch. 11, §61–6.

[30] See Hugh of Saint-Cher, *Théorie de la prophétie et philosophie de la connaissance aux environs de 1233: La contribution d'Hugues de Saint-Cher (MS Douai 434, Question 481)*, ed. Jean-Pierre Torrell (Louvain, 1977).

people's conduct, bring means of correction to bear on it'.[31] Another anonymous Franciscan preacher, probably of north Italy in the same period, contrasted Saint Dominic, who according to him was a prophet, with Elijah, the greatest of the prophets. Elijah had defeated the priests of Baal by miraculous fire from heaven; Dominic defeated the heretics by a miraculous ordeal by fire in which the Albigensian scriptures burned and the Bible was left unscathed.[32] Together these two texts tell us what the word 'prophet' meant when applied to John of Vicenza and his confrères: one who reformed the evils of society and did so by drawing on divine, even miraculous, power.

Here one draws close to what contemporaries meant when they spoke of the friars as 'prophets'.[33] The word was elastic in meaning but indicated above all a making present of divine authority. In the stories that circulated, the gift of prophecy manifested itself most perfectly in a union of several phenomena: visions or ecstasies, a fearless condemnation of vice and sin, and the miraculous defence and vindication of these activities. The power of prophecy allowed the preacher to declare in an authoritative way the will of God in a certain situation, and his word became a definitive judgement, at least until it was reversed by its author,[34] or by a more powerful prophet.[35]

The more theatrical aspects of prophecy, visions for example, are common in the sources. They are often incidental to, or a mere decoration of, the two more important functions—arbitration and correction. Still, visions occasionally receive attention in their own right. A divine revelation could arrive at the most unexpected time and place. The Dominican prior at Brescia, Guala, later bishop of Bergamo, was taking a nap in the bell-tower of his convent when, in a dream, he saw the sky open and two brilliant white ladders descend from heaven. At the top of one, he saw

---

[31] Luke the Lector, *Sermones de Adventu et Festivis*, Biblioteca Antoniana MS 466, fo. 198ʳ: '[P]rophete vocantur quia archana misteriorum Dei contemplant et prout vident mores hominum adhibent modos curationum.'

[32] *Sermones de Tempore et de Sanctis*, fo. 75ᵛ. Origin and date of this manuscript from the new catalogue of the Antoniana: Giuseppe Abate and Giovanni Luisetto, *Codici e manoscritti della Biblioteca Antoniana* (3 vols.; Vicenza, 1975), ii. 461.

[33] The identification of preacher and prophet was not new. It appears as early as the 11th cent. in the vita of the hermit-preacher Bernard of Tiron; see Geoffroy le Gros, *Vita Beati Bernardi* [*Tironiensis*], ed. Godefridus Henschenius (AS 11: Apr. II; Paris, 1866), 225–6, (ch. 2, §14–15).

[34] As, for example, during John's arbitrations at Bologna. John reversed his decision twice, with no effect on his status as prophet. See Sutter, 66–8.

[35] Magli, *Gli uomini de la penitenza*, 37, sees here a parallel with Melanesian prophecy.

Christ, at the top of the other, the Virgin. They were drawing up to heaven a friar seated upon a chair with his cowl pulled over his face, a Dominican burial custom. The ladders were then pulled up and the whole vision disappeared amid the rejoicing of multitudes of angels. When Guala later heard that, at the very time of his vision, Dominic had died at the Friars Preachers' house in Bologna, he realized that his dream was really a vision sent from God.[36] He proceeded to dedicate himself to promoting the cult of the deceased founder of his order, no doubt using this vision as a sermon illustration.

Sometimes visions provided a kind of occupational security, a preacher's insurance policy, so to speak. Called to Brother Elias's death bed, the famous peace-maker Gerard of Modena laboured all day to reconcile the estranged Franciscan minister general to his brethren, but in vain. That night, when he was unable to sleep, Gerard saw demons in the form of bats flying throughout the house and he heard horrible voices and cries. Shortly after, Elias died, unreconciled to the brethren. The vision reassured Gerard that no mediator could have succeeded in making peace between the friars and Elias, since the dying minister had already fallen into the hands of evil powers.[37] Gerard's reputation as peace-maker survived intact.

At other times, the preachers' visions took on meaning only in retrospect. Bartholomew of Trent reports secondhand a story that he received from Giles of Parma, the prior of Faenza. He had seen, in a dream, the Dominican James of Faenza pulling a serpent out of the foundation of the convent under construction in Constance. Later he learned the meaning of the vision. A woman came to James with the report that a wicked man was hiding in the basement of the new building. James went and convinced the man to repent of some shocking, but unnamed, sins involving illicit sex and commerce with the devil. James then confessed him. Outraged demons, now bereft of their prey, put up a horrific fight and nearly destroyed the foundations of the building before the terrified

---

[36] Canonization process for St Dominic according to the text in Quétif-Échard, i. 22. This story is repeated many times in Dominican folklore and art: in the apse fresco above Dominic's tomb by Guido Reni (1615); on the tomb carving by Alessandro Salvolini (1768). These two images show only one ladder, and Dominic's face is uncovered—a remarkable discrepancy with the more ancient (1234) version. Artistic reasons or conformity with the mid- to late 13th-cent. version of the vision in Galvano Fiamma, *Cronica Maior Ordinis Praedicatorum* (ed. in Gundisalvo Odetto 'La Cronaca maggiore dell'Ordine Domenicano di Galvano Fiamma: Frammenti inediti', *AFP* 10 (1940), 345), probably dictated the change.

[37] Salimbene, 105–6.

James managed to exorcise them.[38] Thwarted in destroying the building, the demons had to be satisfied with smiting the repentant sinner with leprosy. Giles asked and received permission from him to recount the story in his sermons.

Like nature-miracles and healing, the gift of prophecy showed a marked tendency to reflect on to the preacher himself and to draw attention to him. Anthony of Padua's earliest recorded prophetic utterance, made to some of his closest friends, was a prediction of his own death.[39] The Dominicans, not to be outdone, reported that during a sermon before large crowds in Romagna, Peter of Verona predicted his own death at the hands of the heretics 'before the coming Easter'. He then went on to predict the turmoil that would afflict the region within a short time.[40] Considering the animosity that his preaching and inquisitorial activity were already generating, Peter probably did not need much prophetic insight to guess at his own end.

The prophetic role, like miracle-working, also risked becoming trivialized, becoming subordinate to putting on a good show. John of Vicenza, in the last recorded incident of his life, took up, for one last time, the mantle of the prophet. In 1262, just as the Friars Preachers were convening in Paris to elect their new master general, John was preaching to the crowds in Bologna according to his usual practice. He held up a blank piece of parchment, showed it to the people and ordered it locked in a safe place. 'Tomorrow you will find on it the name of the new general,' John declared. At which the story-teller interjects, 'he can hardly have been a false prophet, it happened just as he predicted': John of Vercelli was elected general.[41] One might, on occasion, wonder how seriously the preachers' contemporaries viewed some of these antics. There is often a magic-show atmosphere surrounding John of Vicenza's miracles.

Yet, when all was said and done, there was an element of supernatural, almost demonic, power mixed with the theatrics, and this had to be taken seriously. The prophetic office, above all, threw the preacher's

---

[38] Bartholomew of Trent, *Liber Miraculorum Beate Maria Virginis*, Bologna, Biblioteca Universitaria MS 1794, fos. 79ᵛ–80ʳ.

[39] *Assidua* (ch. 14), 346–8.

[40] If we trust Taegio (ch. 2, §15), 698–9, whose original is Pietro Calo. See Dondaine, 'Saint-Pierre-Martyr', 134.

[41] Our source is Leandro Alberti, *De Viris Illustribus Ordinis Praedicatorum* (Ferrara, 1516), fo. 184ᵛ. He tells us his source is a chronicle of Galvano Fiamma. This probably means the *Cronica Major*, the remaining fragments of which lack this text. The passage may be conveniently found in AS 28: Jul. 1, 425: 'Haud falsus vates fuit porro, ut praedixerat inventum est.'

contact with the supernatural world into strong relief.[42] The preacher is endowed with knowledge and authority from on high; his words are to be accepted as those of Christ or the Holy Spirit. This became a common-place in artistic presentations of the preachers. About a century after Ambrose Sansedoni's death, Lorenzo di Pietro, called Il Vecchietta, painted the portrait of him that hangs today in the Pinacoteca of Siena. The artist presented his subject in the act of preaching, holding his mantle with his left hand and gesturing with his right. At his ear is the dove of the Holy Spirit, guiding his words. Although much later in date, there is the same representation of the Holy Spirit imparting divine guidance to John of Vicenza in a fresco at Santa Corona in Vicenza. John too is shown in the act of preaching. Other reports imply that John received the words for his homilies through the ministry of angels. The saintly Bishop of Modena reported to Pope Gregory IX that he had seen angels descend from heaven and guide John as he spoke.[43] The preacher may have been human, but in the eyes of his hearers his words were emphatically from God.

MIRACLES AND AUTHORITY

The office of prophet as exercised during the Devotion is best exemplified by its two components, the authoritative arbitration of disputes and the castigation of sin or vice. Divine intervention in times of conflict or disorder comes to the fore in the earliest description of a homily by Anthony of Padua, that given by the Dominican Bartholomew of Trent, *c.*1233. As Anthony preached during a difficult meeting of the Franciscan

---

[42] Such miracles as these, which validate the preacher or demonstrate his special power and authority, account for 21.6% of the miracles recorded for Alleluia preachers. This is a remarkable statistic. In contrast Sigal, *L'Homme et le miracle*, 291, finds that only 3.6% of miracles worked by 12th-cent. saints while alive had 'glorification d'un saint' as their principal function.

[43] Th. Cant., 424. This story was still current in the 16th cent. See the note written in the margin of an unpublished chronicle by the Dominican Albert of Castello; text in Raymond Creytens, 'Les Écrivains dominicains dans la chronique d'Albert de Castello (1516)', *AFP* 30 (1960), 305–6: 'Fratrem quoque Ioannem Vincentinum, qui cognominatus est sanctus, et totam Lombardiam suis predicationibus commovit. Nam dum predicaret autem visus est angelus domini sepius qui ei loquebatur ad aurem. Aliquando stella fulgida super eum apparebat, aliquando vero crux in aere super eum a cunctis assistentibus videbatur.'

general chapter, Saint Francis himself appeared and blessed the assembled friars.[44] This incident, which occurred during the troubled times following Francis's death, gives us a glimpse of the hopes that the hearers brought to the friars' preaching. During a crisis a sermon could become the place of contact with the supernatural world, and the preacher provided the point of contact. What counted in an Alleluia preacher was, above all, his ability to present himself as the carrier of the certain charisma or power that linked common mortals to the divine will.[45] It is no surprise that the attacks of the devil would be most fierce during homilies.

This was a power that, divine or not, escaped the usual categories of ecclesiastical discipline. A saint like Peter of Verona preached in Florence without permission, 'under the influence of the Holy Spirit'. This was a practice to be admired in saints but not imitated, said one of his eulogists.[46] That the phenomenon of unlicensed preaching was accepted, indeed often rewarded with a licence, is not surprising given the conditions under which the Alleluia preachers were recruited. The hierarchy would gladly take advantage of a proven orthodox preacher, and it made little sense to give out licences to those who lacked the ability to preach or had not yet proved their ability to do so. But unlicensed preaching remained itself irregular.

The Dominicans, from the beginning, struggled with this problem. Dominican constitutions under Jordan of Saxony (master of the order, 1222–37), seeking to nip the problem in the bud, instructed novice masters to look for those who seemed to have the 'grace of preaching' and to recommend them to the prior for a licence to preach.[47] But preaching without a licence continued unabated, and legislation against it reappeared regularly in the *acta* of Dominican general chapters.[48] The problem had no easy solution. In 1249 the Roman Province of the Dominicans forbade the licensing of any but 'proven preachers, from whose preaching there will be no fear of scandal and who will be willingly

---

[44] Text of Bartholomew in AS 23: Jun. III, 198–9.

[45] Magli, *Gli uomini della penitenza*, 122.

[46] Remigio de' Gerolami, *Sermones de Sanctis*, Florence: Biblioteca Nazionale MS Conventi Soppressi, D. I. 937, fos. 153ᵛ–154ʳ: 'Iste [Petrus] ita fuit raptus ut predicandi Florentie nullam licentiam a prelato petebat. Nec mirum, quia prelatus celestis rapiendo movebat. Nec hoc predico ut imitandum sed mirandum, sicut dixit Gregorius de Iohanne baptista, quod fuit doctus a Spiritu Sancto, quia mirabilis Deus in sanctis suis.' · · ·

[47] Dom. Con. (D. 2, c. 17), 74.

[48] e.g. *ACG* 5, for 1234, '[Admonemus] nullus frater predicet aut confessiones audiat sine speciali licencia prioris sui.'

heard by the people'.[49] The Dominican superiors did not want to give approval to preaching by bores or fools. They tried to discover the gifted in the convent, but preaching well in the monastery to small groups of religious did not necessarily promise success before a crowd of townsfolk in the city square. The great preachers, like St Anthony of Padua, were discovered mostly by accident.[50] They escaped neat categories.

The ecclesiastically ambiguous position of the revivalist preachers is best symbolized in a story told of the Dominican Isnard of Chiampo. One night, in Isnard's priory at Pavia, a holy lay brother had a vision, probably in a dream. He saw the people and clergy of Pavia coming to the Dominican house and asking that one of the friars be given to them to be their pope ('papam'). The lay brother told the story to the subprior and the two went immediately to Isnard, certain that this vision could only apply to him. Isnard, immediately reinterpreting the vision as indicating the people's need of a 'father in heaven', fell down on his knees and made a general confession. Soon after, he died.[51] This story both reveals the preacher as a power outside, if not superior to, the normal ecclesiastical authorities and shows how an orthodox preacher might strive to deflect those sectarian tendencies of his admirers that might have rendered his ministry and activity offensive to church leaders.

Beyond giving them the status of prophets, the friars' reputation for thaumaturgy, perhaps through no fault of their own, recommended them for an even higher office. They were, at least in popular parlance, saints. Some preachers considered in this study—Isnard, Leo de' Valvassori, Anthony of Padua, Bartholomew of Vicenza, and Peter of Verona, for example—were formally raised to the altar. Even the very ambiguous John of Vicenza, for all his unpredictability, political manipulations, and bad temper, had a cult in Vicenza that lasted almost to the present day. Others could claim an ambiguous kind of holiness. This holiness often flowed more from the popular image of the wonder-workers than from any genuine piety, and its hollowness gave the revivalist a certain

---

[49] *ACPR*, 11: 'probati predicatores, de quorum predicationes scandalum non timeatur et qui libenter a populo audientur'.

[50] See J. P. Renard, *La Formation et la désignation des prédicateurs au début de l'Ordre des Prêcheurs (1215–1237)* (Fribourg, 1977), 6, on methods used by Dominicans to discover those with the *gratia predicationis*. He concludes that the Dominicans made little effort to train preachers to preach; rather, they sought out the gifted.

[51] Story in *VF*, 227. Copyists understood the dangerous implications of this story: some changed 'papam' to 'patrem'.

insecurity about his own salvation. Salimbene reported the following story firsthand about his acquaintance, Jacopino of Reggio:

To this Jacopino, lying sick in the Convent of Bologna in the infirmary of the Friars Preachers, there appeared brother Gerard of Modena of the order of Friars Minor on the day that he died. He spoke to him in a friendly fashion and said, 'I am in the glory of God, and how quickly is Christ going to call you, so that you may receive the full reward of your labours and live forever with him whom you served!' Having said this brother Gerard disappeared. Jacopino reported to his brethren exactly what he had seen, and they were all glad.[52]

Jacopino died soon after and was buried at Mantua. At the end, with the aid of his old friend and collaborator Gerard, he had vindicated his holiness to himself and others with a vision not unworthy of his best days in the pulpit.

With the reputation for sanctity came its typical medieval accompaniment, the passion for relics. Since the distribution of tangible souvenirs provided not only a means for diffusing a reputation but also the possibility of new miracles and cures, the revivalists did not discourage their followers from collecting mementoes. In this practice John of Vicenza was probably unusual only in the extent to which he promoted his own cult. Whenever John had his tonsure renewed, about once every 3 weeks according to the Dominican constitutions of the time,[53] he saw to it that the hair was saved to be distributed to his followers. On at least one occasion, John was offended that his fellow Dominicans did not save some for themselves.[54] Whatever the Bolognese Dominicans felt about reserving some of John's relics for themselves, they were glad to supply them to others. There was a certain irreverent Florentine Franciscan, Detesalve, who, having heard of this practice, accepted an invitation to the noonday meal with the Dominicans at San Nicola delle Vigne on condition that they give him some of John's under-tunic as a relic. The Dominicans happily obliged him with a piece. Detesalve then excused himself and went out. He tossed it down the conventual latrine and began shouting for help. The unsuspecting Dominicans arrived to help him, sticking their heads down the privy and fishing around for the lost cloth. As they

---

[52] Salimbene, 73: 'Huic fratri Iacobino infirmo existenti in conventu Bononie et in infirmitorio fratrum Predicatorum circa meridiem sedenti et vigilanti apparuit frater Girardus de Mutina ex ordine fratrum Minorum ea die, qua obierat, et familiariter cum eo locutus est dicens: "In gloria Dei sum, ad quam cito vocabit te Christus, ut ab eo laboris tui mercedem plenam recipias, et cum eo semper habitabis, cui devote servisti". His dictis disparuit frater Girardus.'

[53] Dom. Con. (D. 1, c. 20), 205.   [54] Salimbene, 79.

searched in vain, Detesalve collapsed on the floor, convulsing with laughter.[55] Pretensions to sanctity could easily become the butt of jokes. For every scoffing Detesalve, there were many who accepted the revivalists' relics and holiness in good faith. Nevertheless, this holiness was of ambiguous utility. In spite of his objections, devout women constantly harassed Anthony of Padua. Armed with scissors, they fought to cut off bits of his tunic as relics. People counted themselves blessed if they could touch him, and he had recourse to a group of stout young men to act as bodyguards to keep the crowds at bay. On occasion, he had no choice but to flee his admirers and hide in the woods.[56]

John of Vicenza, preaching daily to the crowds, found himself unable to move from place to place. Tired as he was from the great exertion involved, and perpetually hemmed in by people who begged to touch him, tear off some of his habit, or receive his blessing, he begged a local municipal official (*praepositus*) to lend him a horse with which he could escape to the next town. The man confessed that he had no horse except one so fierce that no one could ride it, a horse more savage than any wild beast. Presented with the raging animal, John tamed it with a blessing in the name of Christ, the Prince of Peace. The horse then knelt and humbly allowed John to mount and ride away. When John offered to return it to the owner, the man refused—he could never ride a horse through which God had worked such a miracle. John used that horse in all his travels. Even after he had tamed it, the horse remained savage for all but him. He alone could ride it.[57] The horse was merely confessing what the crowds already knew—the preacher stood outside the natural order of things, emanating power that even brutes could sense. By its miraculous obedience this horse also granted John an exemption from the Dominican constitutions, which forbade the friars to travel on horseback.

DIVINE PROTECTION

Above the head of John of Vicenza in a portrait by Giovanni Speranza, now in the Cappella Sarego of Santa Corona at Vicenza, are the words

---

[55] Ibid.

[56] *Assidua*, 342.

[57] Th. Cant., 424. The literary model for this horse-taming miracle may be Gregory the Great, *Dialogues* (1. 10. 9), ed. Adalbert de Vogüé (Paris, 1979), i, 100–2.

'Timete Deum', 'Fear God.'[58] The words point to a little-recognized theme in Alleluia preaching. This is not the theme of repentance, which has received much attention, but the theme of fear. If the movement of piety in the twelfth and early thirteenth century was, as some would have it, directed towards a 'humanized piety emphasizing friendship, compassion, humanity and affection,'[59] certain aspects of the Alleluia preaching went markedly against this current. Friars' sermons evoke a bitter conflict between good and evil. A friar preached with the angels of God at his side and the Holy Spirit at his ear; he commanded the devotion and respect of the crowds and exerted great political power. There was much about the preachers that made them frightening and dangerous, even when they were not intent on burning heretics 'from the best families of the city', as John of Vicenza did at Verona, or triggering riots against usurers, as he did at Bologna.[60] It fell to the preacher himself to create the highly charged environment in which the hearer made his choice between good and evil.

In part, the success of the revivalist's preaching flowed from his ability to instil fear. Peter of Verona reportedly heard the confession of a boy who had kicked his mother. The preacher severely reprimanded the youth, telling him that a foot that had kicked its mother was worthy to be cut off and destroyed. The youth man, overawed and frightened, returned home and cut off his foot. The horrified mother, having found her son writhing in pain, called for Peter who miraculously restored the severed limb. The fear that brought the boy to confession was itself caused by Peter's preaching.[61] The focus of this story is not the cure, but rather the proper reaction to sins pointed out by the man of God. By the 1240s the same story already circulated under other forms. One version of it, found among the exempla of Bartholomew of Trent, has a pope, guilty of some unidentified sin that involved his hand, expressing his repentance by cutting it off.[62] The wide circulation of stories of this type gives us

---

[58] Although a restoration, this seems to be the correct reading. See Giovanni and Alvise da Schio, *Fra Giovanni da Vicenza a Paquara: 28 agosto 1233–1933* (Schio, 1933), 65. Before the 19th-cent. restoration it read 'Pacem meam do vobis, pacem relinquo vobis' and gave John's date of death as 1260. See Domenico Bortolan, *Santa Corona: Chiesa e convento dei domenicani in Vicenza, memorie storiche* (Vicenza, 1889), 157.

[59] Colin Morris, *The Discovery of the Individual 1050–1200* (London, 1972), 162.

[60] Parisio, p. 8.

[61] Taegio (ch. 3, §20), 700B, originally in Pietro Calo. The story probably dates to the 13th cent., when it circulated under the name of Anthony of Padua. For the textual problem here see Dondaine, 'Saint-Pierre-Martyr', 142–3.

[62] Bartholomew of Trent, *Liber Miraculorum*, fos. 75ᵛ–76ʳ (miracle 163); this story recalls Christ's words in Matt. 5: 30: 'If your right hand causes you to sin, cut it off.'

some idea of the kind of exaggerated acts of repentance that the sermons
of the greater preachers were supposed to produce.

The preachers operated in a context of extreme religious and social
conflict. A successful preaching-campaign could result in the often-
violent suppression of heresy and vice. There is a palpable sense of conflict
in many stories of the friars.[63] There is a combative element, almost a
rivalry, in the use and abuse of the miraculous.[64]

Theobald of Albinga was a Dominican preacher at Bologna in the
1220s and, like the friars of the Alleluia, famous for his peace-making
and miracles. To him appeared a heretic faking a fever, who begged a
blessing and a cure. Theobald replied, 'I ask God, if you have this fever,
to free you of it and, if not, to give it to you.' The man was taken aback
and said, 'Theobald, you are a holy man, how can you say such a thing?
Just bless me and I shall then be healed.' Theobald replied, 'What I have
said, I have said.' The man went out and collapsed on the church steps.
His distraught wife, a Catholic, returned to Theobald and begged for
help. The preacher came, heard the man's repentant confession and
blessed him. Immediately, 'the man was freed from his fever and his
heretical errors'.[65]

According to an admittedly late legend, Anthony of Padua fearlessly
preached before Ezzelino da Romano, rebuking his every vice. The fierce
war-lord, after initial anger, was terrified into repentance by the appearance of a miraculous light encircling the preacher.[66] This story tells us
more about the respect that a miracle-working preacher was supposed to
instil than it does about Anthony and Ezzelino. In a similar story already
recounted, John harangued the Council of Bologna on 16 April 1233.[67]
During his harangue, the appearance over his head of a miraculous
luminous cross put to rest the doubts of the sceptical. By the end of his
address all were in tears of repentance. God had vindicated the preacher
and his political programme. To this day a fresco by Vittorio Bigari at
the church of San Domenico in Bologna commemorates the event.

---

[63] For the hostility inspired by preaching against heresy, see the attempted poisoning of
Anthony of Padua in *Chron. XXIV Gen.*, 124.

[64] A late legend of Anthony of Padua is exemplary. A debate over the real presence was
finally settled by a trial in which the Blessed Sacrament is shown to a mule, which
genuflects in adoration. See *Chron. XXIV Gen.*, 125–6.

[65] *VF*, 225. A virtually identical story is told of Peter of Verona in Taegio (ch. 2, §18),
699C–D. Dondaine, 'Saint-Pierre-Martyr', 141, considers the Peter of Verona version
derivative. Such chastisement-stories probably circulated for other Alleluia preachers as well.

[66] *Chron. XXIV Gen.*, 139.   [67] Recounted in Borselli, *Cron.*, 22.

Such vindication was a warning for all who would ignore God's chosen agent. The Dominican Isnard of Chiampo, who was both fat and a miracle-worker, was preaching in the principal square of a city, probably Pavia. A heretic loudly interrupted his preaching, shouting to the crowds, 'If that barrel over here in front of me rolls over and breaks my leg, then I'll believe that fat friar Isnard is a holy man!' The barrel, with no one touching it, promptly rolled over and broke his leg. He should have known better: Isnard had been showing his power by working exorcisms to convert heretics. A group of heretics had declared that they would convert to Catholicism, if Isnard could exorcise the demon from a certain heretic by the name of Martin. Isnard kissed him and he was immediately freed from possession. The cure here is less important than the friar's display of his own power. The heretics were converted and Martin became a domestic servant in Isnard's priory.[68]

Isnard was not one to be trifled with; his amiable, rotund person was a locus of potentially dangerous powers. Nor was he by any means unique. An unnamed Franciscan from Padua was preaching at Como during the days of the Alleluia. His sermon was interrupted by the noise of workmen who were constructing a tower nearby for a noted local usurer. In disgust the preacher turned to his hearers, many of whom no doubt were in debt to the usurer, and announced that on a certain day the tower would crumble to its foundations and be destroyed. So it came to pass, and all considered it a great miracle.[69] Did some of the debtors who heard the prophecy help the miracle along?

Disconcerting aspects of the preacher's power could come to light in unexpected ways without any action by the preacher—often, it seems, in spite of him. Peter of Verona was seldom free from the heckling of the scoffers and heretics against whom he directed his preaching. He was dragooned, unwillingly it seems, into public debates. On one such occasion Peter had to debate without an opportunity for preparation and found himself face to face with an adversary rumoured to be among the most subtle and dangerous in the ranks of the heretics. Peter retreated to a nearby chapel of the Blessed Virgin and tearfully prayed for assistance. Suddenly he felt a sense of security and spiritual strength. Then, from the image of the Virgin above the altar, he heard a voice, 'I have prayed for you, Peter, that your faith fail not.' He returned to the appointed place for the debate only to discover that the famous proponent of heresy, overcome by anxiety, could hardly stammer out a question, much less an

---

[68] Both stories in *VF*, 227.   [69] Salimbene, 75.

argument. In short, God had smitten him with a case of theological lockjaw. The same topos appears several times in the legend of Peter of Verona.[70] The hagiographer never portrays Peter in the act of debating— his adversary is always struck dumb by divine intervention or by divinely induced fear and anxiety. The power that healed the sick was also the power that humiliated opponents physically and spiritually.

There was a certain Nicholas of Dacia, a student of law in Bologna, who reported on a sermon by John of Vicenza. During his sermon John surveyed the crowds and noticed that nearly all those present had come wearing crowns of roses, no doubt in celebration of the arrival of the famous preacher. He was ashamed, John thundered, to see his flock wearing flowers when their Saviour had worn a crown of thorns. He then pronounced an excommunication—not against the audience, the narrator tells us, but against the wicked roses that had led the People of God into sin. Terrified, the crowd, from oldest to youngest, threw off their rose crowns. None dared thereafter to touch them.[71] Such curses, excommunications, and threats of divine vengeance were common tactics among the preachers; John also cursed the lands and crops of those who ignored his arbitration decrees.[72] What makes the rose-crown story more interesting than such run-of-the-mill juridical cursing is that it did not end with the contrition of the crowds and the excommunication of the flowers. A few days after the fearful homily, there was a marriage during which a young boy carried a rose crown in his hands. One of the party, a cocky teenager ('adolescens superbus'), grabbed it and placed it on his own head. Perhaps it was a dry one remaining from those excommunicated a few days earlier. It immediately burst into flames and his hair was

---

[70] The most ancient version is in *VF*, 237–9; it is recounted in Taegio (ch. 3, §19), 700B–D, a version which is, according to Dondaine, 'Saint-Pierre-Martyr', 141–2, from the very late Antoninus of Florence, *Chronicorum Opus* (3 vols.; Lyon, 1586), iii. 641. Thomas Agni and Calo also repeat the story.

[71] Th. Cant., 424: '[Joannes vidit] quod sertis ex rosis totus fere circumsedebat populus coronatus. Finito ergo sermone, dixit omnibus, pudeat membra fidelis populi rosis aut floribus sub spinato Christi capite coronari. Et addidit: Excommunico serta rosarum, non populum sed serta, quibus ille utitur in peccatum. Destitit ergo populus, a majore usque ad minimum, portare serta: nec fuit aliquis, qui ausus fuit transgredi verba sancti.'

Concerning miraculous chastisement, see Sigal, *L'Homme et le miracle*, 279, who writes, 'On remarque . . . que les actes de désobéissance punis par les saints de leur vivant et après leur mort ne sont pas du tout les mêmes: pendant leur vie, les saints châtient ceux qui ont persévéré dans une mauvaise action malgré leurs admonestations, alors qu'après leur mort, il punissent surtout ceux qui n'ont pas voulu leur rendre les honneurs réclamés.' Not a single punitive miracle worked by an Alleluia preacher punishes any evil act other than lack of respect for the preacher.

[72] Maurisio, 23.

consumed before he could get the crown off his head. 'So the people were thrown into shock; they learned to fear the words of the holy man and, for all his deeds, they blessed God in everything.'[73]

The focus has, till now, been on the image of the preacher in the popular imagination. The picture is not always flattering. The friars practised a kind of revivalism that easily degenerated into a personality cult, and they preached a piety that, because of its highly charged emotionalism and exploitation of fear, easily shaded on one side into dangerous fanaticism, and on the other into transient enthusiasm. It is not surprising that, even among the preachers themselves, preaching could fall into disrepute. Ambrose Sansedoni returns time and time again to the theme of evil preachers and bad preaching. As in the time of St Paul, he says, many of the evils of the Church arise from the abuses of preaching and from wicked preachers.[74] Above all he was perplexed by the crowd's uncanny preference for almost any preacher other than a holy one. Maybe the crowds found the holy ones less entertaining. Perhaps, he suggests, prayer would be more helpful to the Christian life than hearing homilies.[75]

The Alleluia preachers and their imitators bear partial responsibility for a growing disrepute of preaching in the later thirteenth century. Yet during the period of their greatest success the friars enjoyed immense popularity in spite of their somewhat shoddy techniques. This approval was not rooted in their theatricality—that would have given them a very transient and ephemeral celebrity—but rather arose from their activities' substantial benefits to the life of the north Italian cities. The creation of an audience and the cultivation of miraculous powers carried the friars into the world of peace-making and arbitration. Here they proved their substantial worth to their followers. So, before a judgement can be passed on the preachers of the revival, the fruits of their labours—social, moral, and legal—in the cities where they worked must be examined.

---

[73] Th. Cant., 424: 'Conversus ergo populus in stuporem, sancti viri verba timere didicit, et in factis ejus Dominum per omnia benedixit.'

[74] Sansedoni, *Sermones de Tempore* fo. 15[r]. Later he observes that much of the error and heresy found in the Church results from wicked preachers. The vita of Sansedoni also contains implicit criticism of preachers, see A. Thompson, 'Le tentazioni del predicatore nella vita di Ambrogio Sansedoni', *Bollettino di San Domenico*, 67 (1986), 145–50.

[75] Sansedoni, *Sermones de Tempore*, fo. 54[v].

6
The Revivalist as Peace-Maker

THE culmination of the Great Devotion of 1233 was the social ritual of reconciliation and peace-making. The preaching of the Alleluia was itself, above all, the preaching of peace. Writing years later, the Franciscan Salimbene called it, using a phrase that recalled his Joachimite past, 'a time of tranquillity and peace'.[1] Peace did not result automatically from the friars' preaching; rather, they had to construct it through skilful mediation and legal ratification. A friar's peace programme faced many obstacles. If both parties opposed his call, he was usually doomed to failure, unless he could mobilize outside opinion. If the injured party wanted peace, he had to convince the injurer to admit his fault and make some restitution. If the injurer wanted to put an end to the quarrel, he had to convince the injured that there was nothing dishonourable in forgiveness.

Peace-making by preachers was not a new phenomenon in the early thirteenth century. In their peace-making, the friars stand solidly within a tradition extending back to the Movement of the Peace of God in the eleventh century.[2] In their peace-preaching they resemble the hermit-preachers active in France in the late 1000s and early 1100s. Itinerant hermit-preachers there often acted as reconcilers and peace-makers. Robert of Arbrissel would be typical of this phenomenon. It is recorded that he mediated in a dispute between Bishop Ivo of Chartres and a local abbot and, after Ivo's death, reconciled the canons of Chartres and the local count, who had fallen out over the election of a new

---

[1] Salimbene, 70: 'Fuit autem Alleluia quoddam tempus, quod sic in posterum dictum fuit, scilicet tempus quietis et pacis.' The words 'tempus quietis et pacis' parallel Joachim's on the seventh age of the Church. Salimbene, 236–7, describes his own infatuation with Joachim, which began in the 1240s.

[2] For bibliography on the Peace of God, see H. E. J. Cowdrey, 'The Peace and the Truce of God in the Eleventh Century', *Past and Present*, 46 (1970), 42–67; for a general treatment, see Hartmut Hoffmann, *Gottesfriede und Treuga Dei* (Stuttgart, 1963), and the Marxist analysis of B. Töpfer, *Volk und Kirche zur Zeit der beginnenden Gottesfriedensbewegung in Frankreich* (Berlin, 1957).

bishop.[3] Unfortunately, Robert's biographer does not give us any details of the hermit-preacher's methods of reconciliation, beyond the pious reflection that his holiness had driven off the demons of discord that were the cause of the unrest. Nevertheless, one can surmise that Robert combined mediation with the preaching of penance, since his biographer lays special stress on Robert's penitential practices, in particular, his prayers, fasts, and vigils.[4] This combination of penance and reconciliation is an element of continuity between thirteenth-century preaching and that of the early twelfth century.[5] The example of these earlier figures will occasionally serve to illustrate the working of reconciliation by thirteenth-century preachers.

By the high middle ages, Italian peace-making had become much more formal than the mere coming-together of estranged parties in a new bond of friendship. Peace-making was a judicial process; thus, it was inconceivable outside its legal context. One need only look at the early fourteenth-century story of St Francis of Assisi and the wolf of Gubbio in the *Fioretti* to see that even in story-telling peace-making was inseparable from its legal forms. Francis, like a good peace-mediator, listened to the complaints of the townspeople of Gubbio and then sought out the wolf that had been victimizing them. After convincing the wolf to accept his role as mediator, he exacted the equivalent of an oath, in which the wolf promised to accept him as peace-arbiter. Francis then led the wolf into town and addressed the crowd:

'Listen, dear people. Brother Wolf, who is standing here before you, has promised me and pledged his faith that he will make peace with you and will never hurt you if you promise also to provide for his daily needs. And I pledge my faith as bondsman for Brother Wolf that he will faithfully keep this peace pact [*patto della pace*].'
Then all the people who were assembled there promised in a loud voice to feed the wolf regularly.
And Saint Francis said to the wolf before them all: 'And you, Brother Wolf, do you promise to keep this peace pact, that is, not to hurt any human being, animal or other creature?'

---

[3] André de Fontevrault, *Alia Vita Beati Roberti {de Arbrissello}* (ch. 3, §13), ed. Joannes Bollandus (AS 6: Feb. III, 615F), for arbitration between the bishop and abbot; ibid., (ch. 3, §16), 616C, for count and canons.

[4] Ibid., (ch. 3, §15), 616B.

[5] For the link between peace-making and penance-preaching, see Ida Magli, *Gli uomini della penitenza: Lineamenti antropologici del medioevo italiano* (Milan(?), 1967), 90.

The wolf knelt down and bowed its head, and by twisting its body and wagging its tail and ears it clearly showed to everyone that it would keep the pact as it had promised.

And Saint Francis said: 'Brother Wolf, just as you pledged me your faith concerning this promise when we were outside the city gate, I want you to pledge me your faith and promise here before all these people that you will never betray the promise and bond [*promessa e mallevaria*] that I have made for you.'

Then in the presence of all the people the wolf raised its right paw and put it in Saint Francis's hand.[6]

Although this story is a literary creation dating a century after the Alleluia, its image of peace-making—mediation, oath-taking, public ratification, and surety—is strikingly similar to that of the 1230s and before. By the early thirteenth century such forms were already traditional.

Like the redactor of the *Fioretti*, thirteenth-century preachers and their hearers took the legal forms of peace-making for granted. The Alleluia may well have been an extravagant example of large-scale peace-making, but only its intensity set it apart from the continual quest for harmony that characterized urban Italy during the period of popular rule in the communes (*c.*1220–*c.*50). For that reason, it is best to begin the study of the revivalist as peace-maker with the ordinary means of reconciliation available to citizens of the communes.

CONFLICT AND MEDIATION

Feud and civil strife were less a part of life than a way of life in the Veneto and Emilia-Romagna during the 1230s.[7] In the attempt to sup-

---

[6] *I Fioretti di san Francesco*, 1, 21, ed. Paul Sabatier, (Fonti Francescane, 1; Assisi, 1970), i. 1502 (trans. follows that in *St. Francis of Assisi, Writings and Early Biographies: English Omnibus of the Sources for the Life of St. Francis*, ed. Marion A. Habig (Chicago, 1972), 1350–1, with some corrections). I thank Prof. Cindy Polecritti of the University of California, Santa Cruz, for calling my attention to the procedure for peace-making in this text.

[7] On violence in the Italian communes, see *Violence and Civil Disorder*, in particular the essay by the editor Lauro Martines himself, 'Political Violence in the Thirteenth Century', ibid. 331–55. He locates the origins of factional strife in rapid economic change, ibid. 337–8, and a vacuum of legitimate authority, ibid. 341–3. In the jerry-built systems of communal justice and government, Martines, 351–3, sees violence not as evil but as 'necessary' for urban development. On the extreme political disorder of 13th-cent. Romagna, see John Larner, *The Lords of the Romagna: Romagnol Society and the Origins of the Signorie* (Ithaca, NY, 1965), esp. 40–7.

press violence, most cities of the region delegated major responsibility for reconciliation of disputes and keeping of the peace to their chief administrator, the *podestà* or *rector*. By the mid- to late thirteenth century, the oaths of these officials usually bound them to do equal justice to all and to reconcile any conflict that might arise, using force if necessary.[8] In this endeavour, the chief officers acted in concert with other functionaries of the commune.

Although the technicalities of mediation and the means of enforcement elude us for the most part, lines of responsibility for peace-making are clear. The Vicenza statutes of 1265, which are in continuity with those of the 1230s, had the *podestà* recite this oath on entering office:

Also, I shall be held within four months of entering my office, along with the Ancients of the commune of Vicenza or through myself alone, to use all my strength to bring to agreement and settle the feuds, conflicts, animosities, and grudges that I shall have found to exist between any persons of Vincenza and of the district of Vicenza; and I shall investigate any dispute or disputes that arose before my entrance into office that might surface during my tenure and I shall be held, as far as I am able, to reconcile and settle them before leaving office, and I am bound to compel those in conflict to make peace, reconciliation, and settlement.[9]

It should be noted that the *podestà* bound himself not only to arbitrate disputes, but also to compel the disputants to make peace. This would imply, at least in theory, some aggressiveness on his part in putting down unrest and violence. When possible, however, he avoided forceful intervention, since it could merely exacerbate or prolong the violence.

Forums did exist, in particular the ecclesiastical and civil courts, to which disputants could turn. Nevertheless, parties often had reasons to avoid these tribunals. Some disputes, such as that between the commune and

---

[8] On the role of peace-making in 'communal ideology', see Antonio Ivan Pini, *Città, comuni e corporazioni nel medioevo italiano* (Bologna, 1986), 176–8.

[9] Vic. Stat., i. 14–15: 'Item quod tenear una cum anciani communis Vicentie vel per me toto meo posse concordare et sedere infra IIII menses mei regiminis guarras, discordias, malivolencias et odia, que sciero esse inter aliquas personas Vicentie et Vicentini districtus, et hec inquiram officio meo ortas et orta ante introitum mei regiminis que meo tempore orirentur concordare et sedere tenear ante exitum mei regiminis, et quod possim et debeam discordes cogere ad paces concordias, et sedationes faciendas.' The editor of the statutes, Fedele Lampertico, argues for a general continuity of the Vicenza municipal institutions before and after the tyranny of Ezzelino da Romano, 1237–59. The oath of the *podestà* of Verona, dated 1276, is worded in a fashion almost identical to that of his counterpart at Vicenza, except that the *podestà* is to be assisted in his work by a committee of sixteen and a group of notaries. See Ver. 1276 Stat. (ch. 30), i. 39.

bishop of Bologna mediated by John of Vicenza, were over the legal relationship of civil and ecclesiastical authorities themselves. Such conflicts of jurisdiction also figured in several of John's other arbitrations in the Veneto. Even more commonly, the long-standing and violent nature of a feud discouraged recourse to the usual forums. Warring factions had first to agree to negotiate, something their anger and pride might well prevent. To encourage submission of issues to arbitration, the friars had one proven tool that the regular courts lacked—their preaching.

By the fourteenth century, communes had begun to enact laws compelling feuding parties to submit their quarrels to judgement, but, in spite of devices like the Veronese *podestà*'s oath, aggressive communal intervention in private conflicts does not appear typical of the early thirteenth century. Rather, the commune itself engaged outside reconcilers, to avoid municipal involvement in what appeared to them private, clan, or family matters. The 1228 statutes of Verona, for example, provided that the *podestà* seize and punish violent malefactors only if they had not made peace on their own.[10]

In Bologna, where there existed local defence-organizations, the Società delle Armi, peace-making could devolve on those bodies.[11] The Società were in great part internal peacekeeping forces—or better, vigilante groups—and, as such, their membership pledged not only to help maintain the peace but not to break it themselves.[12] When discord arose, they also served as a forum for reconciliation. The statutes of one society, the Traverse di Porta S. Procolo, empowered its directors (*ministrales*) to act as mediators in disputes that arose among the membership.[13] They were to reconcile any quarrels between members over questions of persons or money, and members who refused to accept arbitration were fined a

[10] Ver. 1228 Stat. (ch. 102), 79. Although punishments are provided for homicide, ibid. (ch. 83–5), 65–7, and peace-breaking, ibid. (ch. 86), 67–8, the government is not commissioned to force reconciliations.

[11] As previously noted, these bodies were visible participants in John of Vicenza's processions: Borselli, *Cron.*, 22; Rampona, 102. Bonatti, *De Astronomia Tractatus X*, ed. Nicolaus Prukner (Basel, 1550), fo. 211, also tells us that John went around Bologna accompanied by crowds of armed men, most likely members of the Società delle Armi.

[12] Bol. Armi Stat. (ch. 30), 139, for the Traverse di Porta S. Procolo (dated 1231). The same is the case for the other pre-1233 statutes, those of the Balzani (1230), ibid. (ch. 2), 122–3. Such armed societies were common to all cities that passed under popular rule. For the example of Siena see William Bowsky, *A Medieval Italian Commune: Siena under the Nine, 1287–1355* (Berkeley, Calif., 1981), 127–8.

[13] Unfortunately only two Società statutes dating from before 1233 have been preserved. Similar institutions for reconciliation and peace-making appear in the post-1233 statutes of other societies as well.

minimum of 10s. bolognese.[14] When a dispute arose among the directors themselves, this was to be settled according to the will of a majority of the other directors.[15] It is likely that other semi-public means of arbitration also existed in Bologna by 1233, even if these have left no legal remains.

The success of an arbitration system depended principally on a mediator's ability to impose his will and the willingness of the parties involved to abide by his decisions. Beyond this, even if—and this is far from certain—some mediators could impose their decisions on individual members of the commune, there remained certain varieties of conflict that were not amenable to ordinary means of reconciliation. Such a conflict appears with all its frightful ferocity in the 1276 Verona statutes. These statutes are from a period later than that under consideration, but the conflict they envision would not have been out of place in the 1230s. Following a period of unrest that ended with the exile of the San Bonifacio faction, the opposition, now in control of the commune, enacted a statute to prevent the exiles' return. According to this decree anyone, even the *podestà* himself, who agitated for peace and reconciliation, or even cried 'Peace! Peace!' in the streets, was to be decapitated. If the agitator was a woman, she was to be burned; if a cleric, hanged. Objection to the statute itself brought a stiff fine of £100.[16] Given such legislation, desire for reconciliation, no matter how widespread, could hardly surface inside the commune itself. Return of the San Bonifacio party required either an overthrow of the government or outside intervention.

Impediments to reconciliation, and the reluctance of the communes to involve themselves directly in peace-making, forced individuals and cities to engage outsiders when they felt the need to make peace.[17] The office of *podestà* had itself arisen in this way. This need for independent mediators probably increased after the Fourth Lateran Council forbade the use of the

[14] Bol. Armi Stat. (ch. 22), 138: 'Item statuimus . . . quod ministrales teneantur bona fide concordare omnes discordias que fuerint inter aliquos sotios . . . et quicumque noluerit eiis obedire . . . solvat nomine banni x sol. bon., et plus si placuerit sotietati.'

[15] Bol. Armi Stat. (ch. 25), 139: 'Item statuimus . . . quod, si contigerit quod aliqua discordia oriretur inter ministrales, quod, ubi maior pars ministralium concors fuerit, quod alii teneantur stare.'

[16] Ver. 1276 Stat. (§112), 463.

[17] For anthropological consideration of 'outsider mediation', see G. Kingsley Garbett, 'Spirit Mediums as Mediators in Korekore', in John Beattie *et al.* (ed.), *Spirit Mediumship and Society in Africa* (New York, 1969), esp. 106.

ordeal, with its impersonalizing rituals; now that they could no longer invoke the judgement of God, parties had to find arbitrators to serve as God's surrogate. Besides the *podestà*, others were also acting as arbitrators by the early 1200s. Above all, if the few published peace-instruments for northern Italy in this period are representative, disputants had recourse to religious in their search for mediation.[18] Certain characteristics of the friars, such as their reputation for direct contact with God and their particular role as spiritual advisers to the communal élites, especially recommended them for this office.[19] They had the potential of providing the kind of impersonal and divinely sanctioned judgements previously provided by the ordeal. Also like the ordeal, the friars could provide the ritualized forum in which hostile parties could meet on 'neutral ground'.[20]

Formally speaking, peace-making by friars had much in common with peace-making by secular mediators. The forms and procedures that Gerard of Modena employed at Parma were institutionalized in that city's statutes and provided a framework for later non-religious mediators such as Ghiberto de Gente during his peace-making as *podestà* in 1252.[21] Other Alleluia friars probably followed procedures similar to Gerard's, but evidence for this is lacking. The methods developed in the early

---

[18] For a sample of north Italian peace-instruments of the early 13th cent., see Verci, i, 'Appendice', and Luigi Tonini, *Della storia civile e sacra riminese* (Rimini, 1862), iii, 'Appendice'. The most common arbiters in these disputes are *podestà*, followed by religious, usually friars but occasionally monks. Lay arbiters are less common. I find no regular canons or secular clergy. These latter groups may have seemed too involved in communal affairs. One thing *podestà* and mendicant religious had in common was their status as 'outsiders'.

[19] Donald Weinstein and Rudolf M. Bell, *Saints and Society: The Two Worlds of Western Christendom, 1000–1700* (Chicago, 1982), 149, note the growing tendency in the later middle ages for disputants to call on 'holy' men and women as mediators. They suggest that these were favoured partly for their purported prophetic knowledge of God's will.

[20] On the defusing of disputes by the ordeal, see Peter Brown, 'Society and the Supernatural: A Medieval Change' in *Society and the Holy in Late Antiquity* (Berkeley, Calif., 1982), 313–14.

[21] Parma Stat., 303–4. For Ghiberto de Gente's commissions to arbitrate, see ibid. 209–61; for his decrees, ibid. 217–23. Salimbene, 447, praised him for his peace-making and strengthening of the city's defences: 'Anno Domini M°CCLII° [Salimbene's date is off by one year, see note of editor, Salimbene, 447, n. 4] domnus Ghibertus de Gente, civis Parmensis, assumpsit sibi dominium Parme cum adiutorio beccariorum Parme, quod tenuit multis annis. Et facit in suo dominio duo bona: Unum, quod reduxit cives Parmenses ad pacem. Aliud, quod fecit murari aliquas portas dicte civitatis.' Salimbene, 447–50, lists eight evils which he perpetrated and for which he was deposed; they amount to an attempt to establish a *signoria* hereditary in his family. Salimbene, 451, also tells us that he tried himself to convince the deposed *podestà* to become a Franciscan!

thirteenth century seem to have remained mostly unchanged throughout the later middle ages.[22]

THE VARIETIES OF CONFLICT

The few extant stories of reconciliations mediated by preaching friars, as well as offhand remarks in chronicles about peace-making, suggest that feuds arising from physical violence were the principal, or at least the most prominent, object of the friars' reconciliations.[23] The lengthy homicide-sections of extant thirteenth-century communal statutes suggest the same preoccupation with physical violence. These statutes carefully tabulate schedules of fines and punishments for murder and violence and make provision for the rights of heirs and relatives.

Beyond such tabulations of fines, municipal statutes against homicide seem to share two other characteristics: a concern to prevent the escalation of violence and a tendency to make the prevention and prosecution of violent crime a family responsibility. The lengthy treatment of such crimes in the mid-thirteenth-century statutes of Padua may be taken as typical. There, homicides are to suffer capital punishment ('ultimum supplicium') if they do not make peace with their victim's heirs within a month of the crime.[24] Padua also allowed guardians to initiate actions against those who had killed one of their wards.[25] The last Padua statute on homicide (719) makes the father of a family explicitly responsible for homicides committed by his slaves and unmarried children, thereby reinforcing the family's role in the prevention and punishment of violent crime. The householder's punishment was to be the destruction of his

---

[22] The major change in peace-making during the 200 years after the Alleluia was a growing involvement of communal governments and a greater institutionalization of the role of peace-maker. See e.g. Brucker, *Florentine Politics and Society, 1343–1378* (Princeton, NJ, 1962), 128–9, for the growing use of public power to coerce peace-making in 14th- and 15th-cent. Tuscany.

[23] J. K. Hyde, in his essay, 'Contemporary Views on Faction and Civil Strife in Thirteenth and Fourteenth Century Italy', in Martines (ed.), *Violence and Civil Disorder*, 273–307, reviews the medieval Italian opinions about the causes of unrest. They, starting with Dante, explained it as the result of moral failings, in particular 'Superbia, invidia e avarizia', ibid. 276. The authors of the vita of Ambrose Sansedoni would subscribe to this theory. Today we would more likely identify the origins as economic or political, e.g. Martines, 'Political Violence', 351–3.

[24] Padua Stat. (712–21), 239–42.   [25] Ibid. (715), 240.

dwelling. One remarkable aspect of the Padua laws is that no explicit
distinction is made between voluntary and involuntary homicide. The
impression given is that the laws were enacted as much to prevent the
taking of revenge in the wake of a killing as to punish its perpetrator.
The laws appear to take it for granted that few criminals will actually
suffer the statutory death-penalty. A homicide's failure to appear within
eight days of being summoned was punished by imposition of the ban
and the confiscation of his dwelling (717). It seems hard to believe that
someone guilty of homicide would willingly present himself to receive the
ultimate penalty. Flight, followed by imposition of the ban, must have
been the normal course of events. Nevertheless, the Paduans did impose
the ban on anyone who assisted a fleeing homicide (722). The ban could
be lifted after the payment of a fine of £100 for a knight or £50 for a
foot soldier. Padua also provided for the punishment of reckless acts that
might result in homicide.[26]

Other statutes of the period show marked similarities to those of
Padua.[27] But there are some interesting variations. Vicenza, for example,
also punished homicides with death, but provided for the confiscation of
their goods if they managed to flee the city.[28] Vicenza punished accom-
plices with the ban, removable on payment of a fine of £50 to the commune. Verona's 1228 statutes concerning homicide are similar to those of
Padua and Vicenza, but they do not even bother to prescribe the death
penalty, mandating merely the imposition of the ban and the destruction
of the perpetrator's goods.[29] Verona also gave the city's *rector* the power to
deal with non-premeditated homicides differently from intentional ones
(no particular punishment is specified for the former, but the *rector* must
ensure that amends are made, 'as seems best').[30] Like Verona, Treviso did
not prescribe the death penalty. The Trevisans made do with a fine of
£400 imposed on all homicides over 12 years of age, but they reduced the
fine to £200 if the act was committed by someone inside a tower, e.g. by
throwing out a rock.[31] This strange mitigation may have been intended
for those who killed someone who was attacking their tower.

The overall impression gathered from a reading of the communal

[26] Ibid. (750–2), 252.

[27] e.g. Bol. Stat., i. 316–24; Parma Stat., 278–85.

[28] Vic. Stat., 117–18.

[29] Ver. 1228 Stat. (c. 83–5), 65–7. This legislation is repeated without change in Ver.
1276 Stat. (35–7), 412–13. The statutes of Verona have been systematically studied by
Luigi Simeoni, *Il comune veronese sino ad Ezzelino ed il suo primo statuto* (Miscellanea di storia
Veneta, 3rd ser., 15; Venice, 1922).

[30] Ver. 1228 Stat. (c. 84), 76.   [31] Trev. Stat.. (§308–12), 113–16.

statutes on homicide is that there is a more or less constant tendency to place the prevention of revenge and escalating violence above the direct punishment of crime. All the statutes seem to take it for granted that the perpetrator will flee and that exile (the ban) and confiscation of goods is a sufficient punishment. Little attempt is made to distinguish between voluntary and involuntary homicide—both could set off a feud, and consequently both are equally dangerous to the peace of the commune. The tendency to see the commune's role in homicide as that of discouraging violence in the settlement of what was above all a dispute between families presents itself most strongly in the statutes of Vercelli. There, the commune did not legislate on homicide at all. The commune did, however, enact fairly extensive legislation on peace-breaking, that is, the violation of the formal agreement by which families or individuals had resolved a feud.[32] Communal statutes put a high priority on reducing or removing the hostility between the family of a killer and that of his victim. When a peace-making preacher mediated in a dispute over homicide, he was doing directly what statutory legislation could at best only encourage—removing or reducing a cause of future violence.

Many of the reconciliations orchestrated by the preachers of the Alleluia must have been between families in conflict over an intentional or accidental death. Unfortunately, no mediation preserved in the peace instruments of 1233 deals with murder or physical injury. Reconciliation of a feud, unlike a conflict over land, was unlikely to benefit from a formal registration of the decision. For that reason, legislators rarely enrolled such reconciliations in the statute books. The three surviving reconciliation documents issued for individuals by Gerard of Modena and John of Vicenza all imply a similar cause of dissension: financial or property rights. In the Parma lawbook, Gerard's two statutory reconciliation-decrees, issued for the Hospital of St John and for a bridge corporation, record undefined conflicts, probably financial (tolls or tithes), between these entities and the commune.[33] These acts are so laconic that they record little beyond stipulating the *podestà*'s role in enforcing the friar's decision.

The only extant peace-instrument of John of Vicenza that does not deal with communal politics is a judgement on the possession of a garden. John's verdict, preserved as two statutes in the Bolognese lawbook, first

[32] Vercelli Stat. (§83–5), cols. 1128–9.
[33] Parma Stat., 198, 199.

declares void a contract for the sale by one Armanno of Porta Nuova of a house and garden near the church of San Salvatore in Porta Nuova belonging to the widow and heirs of one Gandulf of Gisso. John orders Armanno to return the money he had received for the sale of the garden and then to surrender the property to the heirs. Then, a separate statute names a certain Juliana as guardian of Gandulf's minor heirs and declares the peace made by her for them with Armanno binding, even after the heirs reach maturity. It directs the *podestà* to enforce this peace with the heavy penalty of £1,000 bol. The exact circumstances of this reconciliation escape us, but it appears intended to put to rest some fraudulent or forced sale for which Gandulf's heirs would later have exacted vengeance. John's decision thus protects a widow and orphans, easy prey to the unscrupulous. A similar concern for family rights and wardship will be seen to motivate several of Gerard's statute reforms at Parma.

Some conflicts arose originally from real or imagined insults or from simple contrariness. Hints of this are already present in the frequent stories of friars' intervention in quarrels between wives and husbands. Gerard of Modena worked to reconcile the divided leadership of his own order during the turbulent rule of Saint Francis's successor, Brother Elias.[34] Grudges such as these, based on bad feeling or misunderstanding, normally left no legal instrument or decree of arbitration.

It would be tempting to consider reconciliations of feud, property disputes, and personal quarrels as examples of 'private' arbitration, in contrast to the friars' 'public' interventions in communal and inter-communal disputes. This would be deceptive. Family feuds and 'private' disputes quickly became 'public' disturbances, since disputants habitually invoked their right to support from other family members or the associations to which they belonged.[35] All disputes were of public concern. When, for example, the *Vitae Fratrum* tells us that the rectors of the city of Bologna sent representatives to the Dominican general chapter of 1233 to request that John of Vicenza be allowed to remain in the city, against the intentions of the master of the order, Jordan of Saxony, 'so that he see the fruit of the seed that he had sowed by his preaching', this fruit was, above all, the reconciliation of personal and family disputes—not his (at that time faltering) effort to reconcile the bishop and the commune.[36]

---

[34] For an example of a preacher reconciling a husband and wife, see Th. Cant., 424; for Gerard's mediation within the Franciscan order, see Salimbene, 105–6 and 162.

[35] A right included, for example, in the oath of the members of the Balzani, Bol. Armi Stat., 122–3.

[36] VF, 138.

Jordan, not wishing to see John's activities restricted to Bologna and perhaps concerned about the revivalist becoming overly enmeshed in local politics, refused the commune's petition.[37] His rejection of the request did not prevent him from later lending his own authority to John's definitive decision in the dispute between the commune and the bishop on 20 June, since in that document John invoked the authority of Jordan as master of the order.[38]

Since political arbitration affected larger arenas and more sensitive issues, the friars' mediation of communal and inter-communal disputes has left more documents than that of private disputes or family controversies. On the communal level one can discern three overlapping areas of arbitration: factional strife, disputes over civil and ecclesiastical jurisdiction, and conflict over feudal rights. Although the friars did mediate in factional disputes, there remain no Alleluia peace-pacts or reconciliations between communal factions. This lack of direct documentation should not deceive us. The presence of general amnesties for the banned or exiled and reports that the preachers regularly released those gaoled for political crimes suggest that some part, if not the greater part, of the friars' mediation concerned civil strife.[39] Examples of the two latter kinds of mediation have survived.

The best-documented example of mediation in a dispute over civil and ecclesiastical jurisdiction, one that also included feudal rights, is that by John of Vicenza between the bishop and commune of Bologna.[40] This dispute, which began in 1231 as a confrontation over tithes, had escalated

---

[37] As we know, the pope was also concerned that John be able to leave Bologna. We know this from his letter of 29 Apr. 1233 to the commune of Bologna. Text in Savioli, iii. 125–6 and *BOP* i. 48, letter 74 (Auvray, i. 713, letter 1268). The pope also wrote to John at the same time asking him to come to Tuscany as mediator between Siena and Florence, *BOP* i. 48 (Auvray, i. 713, letter 1270).

[38] Savioli, i. 132: 'Et hanc sententiam laudo, arbitror et precipio et promulgo de speciali mandato et licentia fratris Jordani Magistri ordinis fratrum Predicatorum.' Would it be too much to suggest that this phrase indicates that, although Jordan first directed John to comply with Pope Gregory's request, he later acceded to the commune's request and allowed John to finish his mediation between bishop and commune and declare a first decision on 31 May and then a second on 20 June? This would certainly explain why Jordan's approbation is present on this decree but not on John's commission to arbitrate nor on his Paquara decree.

[39] We are told by Borselli, *Cron.*, 22, and Rampona, 101, that John freed those imprisoned in Bologna. He also ordered political prisoners released in the cities of the Veneto, see Maurisio, 31–2, and Godi, 10. See also the amnesty statutes of Gerard in Parma Stat., 304–5.

[40] Along with narrative references to the arbitration, we possess the texts of John's compromis of 19 Apr. 1233, in Savioli, iii. 123–5, the revision of 29 Apr. 1233, Savioli, iii. 125–6, and the final version of the treaty, that of 20 Jun. 1233, Savioli, iii. 126–33.

to a dispute over criminal jurisdiction after an episcopal investigation of a crime at San Giovanni in Persiceto, a district over which the *podestà*, Frederick of Lavellongo, claimed jurisdiction for the commune. The commune then seized several castles pertaining to the bishopric in retaliation, and the bishop declared an interdict and left the city for Reggio.[41] Two bishops representing the pope then excommunicated the *podestà* and his ministers. By 1233, when John arrived at Bologna, the dispute was already of long standing. As was probably the case in most communal conflicts, this jurisdictional dispute concerned the bishop's feudal rights of justice and their erosion by the commune.

John delivered his second and final judgement in this dispute on 20 June 1233. It can serve as a model for arbitration decrees of the period. This instrument is lengthy but its contents fall into three clear sections. First John treats the legal question of rights: the commune has the right to punish malefactors resident on the bishop's lands who harm those living outside of them. It grants jurisdiction to the commune over certain crimes that are especially contrary to public order, such as murder, mutilation, arson, perjury, fraud, use of false measures, and counterfeiting, even if these occur on episcopal lands.[42] It then clarifies the procedure for settling questions of conflicting jurisdiction. Finally, John rules on the factual question of which lands belong to the bishopric and appends a list of those lands to his judgement.[43]

Comprehensively treating the areas of law, fact, and procedure, this declaration may be taken as an example of the most comprehensive sort of arbitration decree issued by friars in their mediation. Solemn and comprehensive as it is, the decree is above all concrete. John does not explicitly cite any theoretical principles from feudal custom, canon, or Roman law. The principles behind the judgement are found in the traditional practice of the north Italian communes: John carefully respects established custom, such as a commune's right to administer justice in certain grave cases such as murder, arson, mutilation, and counterfeiting anywhere within its district. His settlement is protected by the establishment of a means to resolve any future disputes. Finally, the decree became an item of public record, both by a vocal proclamation in the city's council and by the attestation of witnesses and a notary, Bolognitto of Strada Maggiore.[44]

---

[41] For the background of this dispute see Sutter, 63–4, and Ch. 2 of this work.

[42] Savioli, iii. 128–9.   [43] Savioli, iii. 129, list in Savioli, iii. 132–3.

[44] Although I do not find any other arbitrations between a commune and a bishopric by an Alleluia preacher, there is a reconciliation decree for the bishop of Parma and his

The thirteenth century saw other reconciliations at the intra-communal level—for example, that attempted by the chronicler Salimbene between factions at Modena in 1265.[45] Peter of Verona organized peace-making at Florence and negotiated pacts between various cities of the Romagna. Jordan of Saxony did the same during a dispute between Padua and Venice. Unfortunately, no documents remain to show how these reconciliations were accomplished.[46]

The *Legend of Perugia* provides one incident where the peace-maker's personal actions in a communal dispute are recorded. It records St Francis of Assisi's reconciliation of the bishop and *podestà* of his home town.[47] The exact cause of the two leaders' falling out is never explained, but the conflict had quickly escalated to the point where the bishop had excommunicated the *podestà* and the *podestà* had forbidden citizens to have any commercial dealings with the bishop. Francis, who was sick with his final illness at the time, did not intervene directly. Rather, he composed a new stanza for his 'Canticle of the Sun', praising God for those who forgive wrongs and make peace. He then had a companion sing the canticle in the presence of the two men. The performance seems to have been sufficient to create an environment for reconciliation, since we are told that the *podestà* threw himself at the bishop's feet and begged forgiveness in the name of God and Francis. The bishop immediately responded by humbly apologizing for his short temper. The two then exchanged a kiss of peace. Those present took it to be a great miracle that the two had reconciled without any 'reference to, or exchange on, the quarrel'.[48] The omission of negotiation certainly was remarkable, if not

commune, dated 10 Jul. 1221; see Parma Stat., 194–8. This dispute began in 1219 over the right of the bishop to invest the *podestà* and quickly escalated to a battle over taxation and rights of justice. In the absence of a means of arbitration, this dispute became violent and finally resulted in excommunications, not only by the bishop, but also by the bishops of Ostia and Velletri in the name of the pope. Reconciliation occurred only after the naming of a new *podestà* in 1221. Had reconciliation not been available, the dispute in Bologna might have followed a similar course.

[45] Salimbene, 391.

[46] For peace-making by Peter of Verona at Florence in 1244–5, see Gilles-Gérard Meersseman, 'Les Confréries de Saint-Pierre-Martyr', *AFP* 21 (1951), 51–196. For his reconciliations in the Romagna see Tonini, iii. 528–32; for Jordan of Saxony, see *Archivio Sartori* (Padua, 1984), i. 987.

[47] Recounted in the Anonymus Perusinus, *Legenda*, ch. 44, ed. in *Scripta Leonis, Rufini et Angeli, Sociorum s. Francisci*, ed. Rosalind B. Brooke (Oxford, 1970), 166–70; and *Speculum Perfectionis*, 9. 101, ed. in *Le Speculum Perfectionis; ou, Mémoires de frère Léon sur la seconde partie de la vie de Saint François d'Assise*, ed. Paul Sabatier (2 vols.; Manchester, 1928), i. 291–5.

[48] Anonymus Perusinus, ch. 44, 170: 'pro magno miraculo . . . quod tam cito visitavit eos Dominus et quod sine recordatione alicuius verbi de tanto scandalo ad tantam concordiam redierunt'.

unique, among recorded thirteenth-century reconciliations. Normally the settling of a feud required a careful mediation to set aright the wrongs committed during the dispute and to remove the issues that had triggered it. Perhaps in this case the quarrel was perpetuated by nothing more than the bad temper and spitefulness of the parties. Reconciliations orchestrated by revivalists of the Alleluia and other preachers like them invariably required something more than a poetic gesture to bring about mutual forgiveness.

THE UNION OF PREACHING AND PEACE-MAKING

The author of the *Legend of Perugia* was fully aware that reconciliation usually required some active personal involvement by the peace-maker.[49] He reports that St Francis had dispatched his follower Silvester to Arezzo to proclaim a prayer of exorcism before the gates of the city against the demons that had stirred up factional violence there. And 'it came about by divine mercy and the prayer of the blessed Francis that without any preaching they returned shortly to peace and unity'.[50] The narrator thought it worthy of remark that this peace-making was accomplished without the customary prerequisites of preaching and mediation. In the author's mind such activities were a necessary part of any pacification. This impression is confirmed by the extant narrative descriptions of thirteenth-century peace-making.

Unfortunately, no description of a reconciliation worked by a revivalist of the Alleluia has been preserved. There exist, however, two nearly contemporary descriptions of peace-making by Italian Dominicans. These stories transport us into the world of the Great Devotion. The first appears in the vita of the preacher Ambrose Sansedoni;[51] the second, concerning Theobald of Albinga, which I shall treat in the next chapter, appears in the *Vitae Fratrum* of the Dominicans. Both narratives deal with feuds. Theobald's deals specifically with a blood feud, one that had arisen from a homicide.[52] One important aspect distinguishing peace-making by the revivalists from that by seculars shines through in the story from the life of Ambrose Sansedoni. The preachers placed the reconciliation of

---

[49] Ibid. ch. 81, 228–30.
[50] Ibid. 230–1: 'Et factum est divina miseratione et oratione beati Francisci, quod sine aliqua predicatione paulo post reversi sunt ad pacem et unitatem.'
[51] Alessandrino, 190.   [52] VF, 225.

enemies in an explicitly religious context. Ambrose's biographer tells us that he regularly preached against the lesser classes' desire for what belonged to the greater; he called it a form of idolatry. Among the species of idolatry, Ambrose gave pride of place to one particular vice, the taking of vengeance—because the avenger usurped a prerogative proper to God alone.[53]

On one occasion, a miracle showed God's approval of this message. In Ambrose's own city of Siena there lived a certain man whose lust for revenge had made him deaf to every plea for forgiveness and closed him to any reconciliation with his enemies. He rebuffed all arguments that he make peace, even those of Ambrose himself. After an encounter with the peace-breaker, the frustrated friar took his leave, promising to remember the man at prayer. 'I could not care less whether anyone prays for me!' the man retorted.[54] Despairing of human means, Ambrose prayed that God himself do justice and prevent the man from taking revenge. The peace-making friar did not stop there; he composed a prayer for the sinner's conversion and taught it to the Sienese during a solemn public sermon. At nearly the same hour the stubborn man was publicly proclaiming to his friends and neighbours that he would never make peace. He went even further, egging them all on to vengeance and urging that they turn a deaf ear to any talk of reconciliation. His personal quarrel was threatening to become a feud.

Meanwhile, Ambrose and the crowd in the Piazza del Campo were praying, and this prayer was so effective that the man's heart suddenly changed. He and his friends arrived before the preacher and now proclaimed themselves amenable to reconciliation. The man knew well what the crowd now expected of him; he plunged into deep mourning and fasting, hardly sleeping for two days. Finally, with his friends and enemies: 'He came to the man of God, begging that he make peace between them and grant them forgiveness for their error. Then the saint, with joy and giving thanks to God, arranged, with great facility, an accord between them.'[55] We see an obvious progression. Ambrose

[53] Alessandrino, 190: 'Dicebat B. Ambrosius in praedicationibus suis vindictam idolatriae peccatum fore; quia ad solum Deum vindicta pertinet: unde homo vindictam faciens, qui Dei sunt occupat, et ideo vindictam faciens maxime dolore debet et poenitentiam facere. Ex hac conclusione inferebat, magnum fore peccatum inferiorum, ambitiose aspirantium ad ea quae juste conveniunt superioribus.'

[54] 'Non curo ut quis oret pro me.'

[55] 'Conferens autem cum amicis et ad pacem eos adhortans, venerunt pariter ad virum Dei, rogantes ut pacem inter eos componeret, et de errore suo veniam daret: de quo beatus, vir admodum laetus effectus, Deo gratias agens magna cum facilitate pacem inter eos composuit.'

preached against vengeance; then he organized the prayer of the people during a sermon; and finally the trouble-maker approached him, probably in public during a homily, and proclaimed his willingness to be reconciled and make peace. In other words, preaching led to the condemnation of division, and this led to the enactment of public rituals to remove it. Other witnesses emphasize the use of sermons as the context for peace-making. The chronicler Rolandino of Padua, who reports extensively on the revival of 1233, tells us that the time and place of John's peace assembly at Paquara were announced during a Solemn Preaching and that the assembly itself began with a sermon on John 14: 27—'My peace I give you, my peace I leave to you.'[56] The earliest life of St Anthony of Padua also connects peace-making and sermonizing:

He called the quarrelling to fraternal peace; he gave liberty to those oppressed by captivity; he ordered returned all property taken by usury or violence, so much so that, having bound their homes and fields as a pledge, people laid the money before his feet and, according to his counsel, made restitution to those whom they had despoiled, without having to be begged or bribed.[57]

Again the same pattern: preaching on peace followed by reconciliation and the removal of the causes of division—here, extortion from the poor.

This combination of preaching and reconciliation seems to have been traditional; the friars were not innovators in this.[58] At the turn of the thirteenth century, Foulques of Neuilly preached against injustice, organizing restitution and reconciliation.[59] Before Foulques, the hermit-

---

[56] Maurisio, 32, confirms this: 'Facta predicacione, pacem pronunciavit et laudavit inter omnes et ordinavit tunc matrimonium inter dominam Adeleitam, filiam domini Alberici de Romano, et Raynaldum, filium Marchionis hestensis.' As does the *Liber Regiminum Padue*, ed. Antonio Bonardi (RIS² 8:1; Città di Castello, 1905), 310.

[57] *Assidua*, 344: 'Discordantes ad fraternam pacem revocabat; captivitate pressos libertati donabat; usuras ac violentas predationes restitui faciebat, in tantum ut, pignori obligatis domibus et agris, ante pedes eius precium ponerent et, consilio ipsius, ablata quoque prece vel precio spoliatis restituerent.' A briefer version of the same report is found in Thomas of Pavia, *Dialogus de Gestis Sanctorum Fratrum Minorum*, ed. Ferdinand M. Delorme (Bibliotheca Franciscana Ascetica Medii Aevi, 5; Quaracchi, 1923), 19–20.

[58] It would be interesting to know how the preachers' role as confessor facilitated the work of peace-making. Confessions seem always to have accompanied the sermons of the itinerant preachers; see, for example, *Assidua*, 346, where we are told that Anthony and a whole team of priests confessed the hearers after his preaching; or, earlier, the regular confessing of his hearers by Bernard of Tiron described in Geoffroy le Gros, *Vita Beati Bernardi {Tironiensis}*, ed. Godefridus Henschenius (AS 11: Apr. II; Paris, 1866) (§59), 236C–D.

[59] Jacques de Vitry, *The* Historia Occidentalis *of Jacques de Vitry: A Critical Edition*, ed. J. F. Hinnebusch (Spicilegium Friburgense, 17; Fribourg, 1972), 98. Here the oppressors seem to have been grain merchants. His preaching appeared to have results: 'Ex quo factum est quod secundum eius verba, pro modico pretio ubique habebantur victualia.'

preachers of the early twelfth century united reconciliation and preaching. In the earliest clear parallel to the patterns seen during the Great Devotion that I have found, Norbert of Xanten used a sermon as the occasion for organizing reconciliations at Monstier near Namur.[60] There, after saying Mass for the Dead—his practice before preaching reconciliation— Norbert announced that when Jesus first sent disciples out to preach he told them that their first words to those they met were to be ' "Peace to this house," and if there were a man of peace there that peace will rest on him.' Invoking this commission he ordered all to leave and be reconciled to their enemies. 'And what then? Adversary factions went from both sides out to the vestibule of the church and then, with relics placed in their midst, after a short time, having forsworn discord and come to accord, they sealed the peace with an oath.'[61]

The forum of the sermon gave the friar certain advantages over a secular mediator. He had at hand apt means to organize public opinion and bring it to bear on the uncooperative. One can sense the social pressure brought to bear on Norbert's hearers, but it is gentle in comparison with the pressure the Sienese peace-breaker must have felt when he heard that a large part of the population of Siena had massed in the Piazza del Campo, praying for his conversion. Ambrose had used his preaching to mobilize public opinion.

Ambrose's biographer tells us that the Sienese peace-breaker immediately entered a period of penance. He knew the proper response to a call for reconciliation. Peace-making was by its essence penitential. Some, for example Vito Fumagalli, have sought to separate penance and peace-making in revivalistic preaching, making this the grounds for speaking of two distinct revivals in 1233. He would have the penitential aspects of the Devotion of 1233 to be something added by the Dominicans to the 'Alleluia', which had essentially been a joyful celebration of religious enthusiasm and peace.[62] This distinction seems untenable. Salimbene did

---

[60] The sources give us no indication of when this preaching took place or whether it was part of a larger peace-preaching campaign.

[61] *Vita Norberti B*, 818A–B: 'Quid plura? Egressa est utraque pars adversaria hinc et inde foras in atrium, reliquis in medio positis; et parvo intervallo facto, adjurata est discordia, et concordia facta, et pax sacramento confirmata.' Other examples of peace-making by Norbert are recorded near Brabant, where a noble who refused the peace meets his sudden death, ibid. 818C–E; and near Couroy, where a peace-breaker trying to leave unreconciled is brought to conversion when his horse miraculously refuses to move, ibid. 818F–819A. The divine interventions in these incidents put one in mind of John of Vicenza.

[62] Fumagalli, 269–70: 'grandioso movimento [di] svariate forme di celebrazione della gioia'. My rejection of this distinction does not, as I noted in Ch. 2, exclude the possibility that Benedict's revival was more lay-dominated than what followed. But the most recent

define the Alleluia as a 'tempus quietis et pacis'. Nevertheless, his descriptions of the joyful 'time of peace' in Parma during the preaching of Benedict Cornetto sound virtually identical to those of the Dominican John of Vicenza's preaching at Faenza and elsewhere.[63] The union of peace and penance is not in itself surprising, since sacramental penance itself originated in rituals of reconciliation to God and neighbour. Norbert's peace-preaching even employed the text of Matthew 5: 23 that made this connection: 'If your brother has something against you, go and be reconciled to your brother and then offer your gift at the altar.'

This traditional union of peace-making and repentance throws the penitential features in Alleluia preaching into greater relief. John of Vicenza, two days before his solemn address to the council of Bologna on 16 May 1233, had the entire populace march barefoot in procession through the city. This procession was conducted with sobs of repentance.[64] His address to the council met with further outbursts of tears.[65] This procession and address were part of the preparations for John's first ruling in the reconciliation of the bishop and commune on the last day of May.[66] The penitential aspect of peace-making surfaced again during John's assembly at Paquara, where 'the great majority was barefoot'.[67] Acts of penitence allowed those being reconciled to surrender the rights

direct treatment of 1233 does not distinguish between two revivals or styles of revival during that year, Daniel A. Brown, 'The Alleluia: A Thirteenth Century Peace Movement', *Archivum Franciscanum Historicum*, 81 (1988), 13.

[63] Salimbene, 70: 'Et vidi, quod in civitate mea Parmensi quelibet vicinia volebat habere vexillum suum occasione processionum. . . . Sic etiam veniebant de villis ad civitatem cum vexillis et societatibus magnis viri et mulieres, pueri et puelle, ut predicationes audirent et Deum laudarent.' Compare this with the description of John in *Chronicon Faventinum*: '[I]verunt Faventini omnes cum pueris et puellis, senes cum iunioribus atque mulieribus, apud Castrum Sancti Petri de Bononia, tam de civitate quam de districtu, cum vexillis omnibus, cruces desuper portantibus, ad predicationem fratris Iohannis.' (*Chronicon Faventinum Magistri Tolosani {AA 20 av. C.-1236}*, ed. Giuseppe Rossini (RIS² 28; Bologna, 1939)).

[64] Borselli, *Cron.*, 22: 'Die 14 madii processionem universalem fieri fecit, ut omnes venirent pedibus nudis; processio celebrata est cum lacrimis multis.' And Rampona, 102: 'Et adì 14 de mazo fu facto procesione dal dicto fra Zohanne cum lo puovolo de Bologna per tucta la città cum li piedi descalzi.' Briefer reports are found in CCB, 101.

[65] Borselli, *Cron.*, 22: 'Ex dulcedine verborum eius multi plorabant.' That this address occurred two days after the penance procession is recorded by Rampona, 101.

[66] On this decision see Sutter, 92–3.

[67] Maurisio, 31: '[P]resbiteri quoque et clerici, milites et populares omnes fere istarum civitatum inermes et cum vexillis cruciatis fuerunt ibidem; tanta fuit ibi innumerosa multitudo fuerunt congregati in unum et in reverenciam ipsius pro maiori parte fuerant discalciati.' This text beautifully reflects the reconciliation of groupings within the society— priests-clerics; knights-citizens—and their ceasing of hostility—even the knights are 'unarmed'. Godi, 10, emphasizes the penitential aspect, 'fere cunctis discalciatis'.

they had previously claimed without losing face, since the capitulation was no longer an act of weakness but a response to God. Compromise with an enemy became a mark not of humiliation but of obedience to the friar's decision as a representative of Christ.

The rituals of penance were, by their nature, marvellously adapted to the creation of concord. Like Ambrose's public prayer for peace-breakers, processions and public acts of penance provided the preacher with a forum for marshalling and displaying public opinion. Processions drew together contrasting groupings in the society, mixing, if only ritually, young and old, men and women, city dwellers and country folk, rich and poor.[68] Such mixed groups, which embraced hostile and opposing factions, then went on foot, often unshod, to some distant place, usually far from the city.[69] The very act of forming companies and marching together was already a commitment to hear out the message of the peace-preacher.[70] It then remained to the preacher to exploit the effects of these collective, community-forming, experiences. Norbert of Xanten gives us a crafty example of how this was done in an earlier age. He once kept a large crowd waiting several hours to hear him preach. Hidden in a wayside chapel, he spent the time in prayer. The frustrated crowds slowly dispersed in disgust to the local taverns, from which they had to come running when it was announced that Norbert had left the chapel and decided to give his sermon.[71] After preaching and reconciliation, the members of the commune could return to the city renewed, and, united by the shared experience of anticipation, undergo repentance and seek resolution of their differences by the preacher. Were there a greater

[68] Note the descriptions in, for example, Salimbene, 71; ibid. 73; and *Chronicon Faventinum*, 159.

[69] Paquara was an open field on the Adige outside of Verona; the people of Faenza marched to Borgo San Pietro, a good distance outside the city; finally Salimbene, 76–7, says that Jacopino saw in his vision John of Vicenza preaching to the Bolognese on the banks of the Reno—well outside the city walls.

[70] The phenomenon of 'liminality' described by Victor Turner, *Dramas, Fields, and Metaphors: Symbolic Action in Human Society* (Ithaca, NY, 1974), seems useful for understanding many aspects of the Alleluia preaching. The creation of 'normative community' can be facilitated by the creation of 'existential community', that is, by the breaking down of the ordinary divisions within a society through promiscuous mixing of its members or by their undertaking of collective ritual acts which reverse or destroy normal roles. Victor and Edith Turner, *Image and Pilgrimage in Christian Culture: Anthropological Perspectives* (New York, 1978), 241, see this as typical of pilgrimages and of revivalistic preaching in general. He writes of American Protestant camp-meetings: 'Significantly, however, the meetings are held not in the midst of a town or village but in a campsite in the countryside, a little apart from a number of villages or homesteads.'

[71] *Vita Norberti B*, (§32–3), 817F–818C.

number of reports, the preachers of the Devotion would certainly be found to have employed techniques similar to those used by Norbert and Ambrose.

It is ironic that the actions by which the friars softened up opposition to reconciliation and mobilized opinion to support it must be reconstructed from fragmentary reports. To the spectators they must have been the most visible and obvious elements of peace-making. For them we have, at least, some fragmentary reports. The other side of reconciliation, its formal legal processes, which contemporary narrative descriptions do not even hint at, remained invisible to the outside observer. For this hidden aspect of reconciliation the evidence is full and ample, but must be sought outside the chronicles and narrative sources. The evidence left by the legal forms and procedures that comprised this invisible but essential part of all medieval reconciliations is sufficient to reconstruct the real heart of a thirteenth-century peace: the peace negotiations and the contract that sealed it.

7
The Revivalist as Arbiter

REPENTANCE and good will alone could not remedy conflicts arising from violence, crime, property disputes, oppression of the defenceless, and transgression of legal or traditional boundaries. Peace-making required more than the social process in which the preacher organized public opinion, nurtured a healthy environment for reconciliation, and moved his hearers to put away their differences and make peace. Legal procedures were necessary, both to ratify the peace and to provide restitution for harm done.[1] Canon law provided little legislation that would provide a forum for peace-making. Gratian himself had undermined the older Church legislation concerning the Peace and Truce of God by justifying defensive war during Lent while leaving intact the protection of non-combatants, but the Third Lateran Council had reaffirmed the traditional bans.[2] But this provided little structure for ending feuds. Nor is there evidence of an organized imposition of the Peace in the cities of north Italy such as has been described for southern France and Catalonia.[3] Peace-making in north Italy during the first half of the thirteenth century seems above all to have been a project for individually commissioned mediators, even if the Church and the commune did what they could to help them and enforce their decisions.[4]

Peace mediation by the friars might be expected to exhibit an *ad hoc* or even informal quality, but this is not the case. The reconciliations and the

---

[1] Literature on medieval arbitration and mediation is reviewed in Thomas Kuehn, 'Arbitration and Law in Renaissance Florence', *Renaissance and Reformation*, 23 (1987), 289–319. This essay and the review of literature it contains shows how much work needs to be done on the topic. Kuehn, 291, thinks that the office of *arbitrator* only appeared in the 1290s with the jurist Guglielmo Durante. The most interesting part of this study is the treatment of particular cases, ibid. 299–311.

[2] On protection of non-combatants, C. 24 q. 3 cc. 22–5; on war in Lent, C. 23 q. 8 c. 15, cf. X 1. 34. 1.

[3] Thomas N. Bisson, 'The Organized Peace in Southern France and Catalonia, *c*.1140–*c*.1233', *American Historical Review*, 82 (1977), 290–311.

[4] In legal terms, then, there seems to be little relationship between the peace-making under study and that during the Movement of the Peace in the 1000s and 1100s, although some broader similarities, as suggested by Dickson may well exist: Gary Dickson, 'The Flagellants of 1260 and the Crusades', *Journal of Medieval History*, 15 (1989), 246.

written peace-agreements they produced were formal legal acts according
to set forms. They had enforceable consequences. In short, the goal of
reconciliation was not simply amicable concord between the parties but a
legally binding contract to observe the terms of peace.

THE PUBLIC FORUM[5]

One of the narrative descriptions of peace-making that has come down to
us throws into relief the political manœuvering and formal procedures
necessary to draft the final peace-agreement.[6] The Dominican Theobald of
Albinga reportedly had a special grace for bringing about reconciliations.
One day, after publicly reconciling a large number of people, he saw to
one side the brother of a murdered man and on the other side the
murderer. He summoned the murderer, presented him to the dead man's
brother and begged him, for the sake of God, to 'give him the peace'
(*redderet pacem*), that is to say, to give him a ritual kiss. But the man, as if
he suddenly saw his dead brother, exploded with rage and began to
threaten the murderer with words and gestures. Theobald invoked God
and said:

'I order you, in the name of Almighty God who made heaven and earth, who
suffered on the cross for us and forgave those who crucified him and prayed for
them, to make peace with this man before you move from this spot.' Marvel of
marvels! The man was then unable to move his feet until he complied.[7]

But another brother of the dead man, less impressed than the hagiographer by miraculous paralysis, on witnessing this reconciliation himself, flew into a rage. Theobald turned to him and commanded him to

---

[5] Interesting parallels to our friars' arbitration may be found in Peter Brown, 'Rise and
Function of the Holy Man in Late Antiquity', *Journal of Roman Studies*, 61 (1971), 80–101,
and his 'Town, Village and Holy Man: The Case of Syria', in *Assimilation et résistance à la
culture gréco-romaine dans le monde ancien* (Paris, 1976), 213–20. Another suggestive study on
the same theme is Evelyne Patlagean, 'Sainteté et pouvoir', in Sergei Hackel (ed.), *The
Byzantine Saint* (University of Birmingham Fourteenth Symposium of Byzantine Studies;
London, 1981), 88–105.

[6] *VF*, 225–6. This friar, whose name appears in *VF* as 'Robaldus' and in other sources
as 'Tibalt', was of the Dominican convent of Milan and was active in Bologna about the year
1220, the year in which he received the Dominican habit from St Dominic himself.

[7] *VF*, 225: ' "Precipio tibi in nomine Dei omnipotentis qui fecit celum et terram, qui in
cruce pro nobis passus est et suis crucifixoribus pepercit et pro eis oravit, quod antequam
pedes moveas, cum isto facias pacem." Mira res! Pedes movere non potuit.'

calm his anger, take the murderer home to dinner, and return the next day so that he could draw up a peace instrument (*instrumentum pacis*) between them. 'Thereupon they completely fulfilled everything that the servant of God had commanded.'[8] Under the peculiar twists of this story, the process of peace-making moving fitfully forward and finally yielding its legally enforceable results is distinctly visible.

What can legal and statutory evidence add to the knowledge of the process outlined in this narrative? The first formal step to reconciliation (one that is not visible in the Theobald story), after the preacher's initial canvassing of opinion and sermon on repentance and forgiveness of enemies, was his clearing the way for disputants to lay their quarrels at his feet for mediation. Beyond moral encouragement and private or public pressure, several concrete steps had to be taken to allow negotiations to proceed. Violent feuds and other crimes normally led to the imposition of assorted punishments and penalties. These had to be relieved. John of Vicenza ordered the freeing of those Bolognese held in prison. The Rampona chronicle sweepingly says that he 'had all the prisoners in the gaols of Bologna released'.[9] Borselli, a later but well-informed witness, restricts this release to those gaoled for debt (*propter debita incarceratos*) and links the action to his amnesty for debtors. Some of their debts were, most likely, the result of usurious loans.[10] Since Borselli places debt relief among John's actions against usurers, he may have intentionally restricted the limits of the amnesty; Rampona is right when he extends the relief to those imprisoned for other reasons as well. But their release required certain preliminaries that the release of debtors did not.

In practice, the customary punishment for violent or political crime was exile and the imposition of the ban. Preparation for peace-making dictated that those under the ban receive relief. The mediator could give this legal form by publishing norms of amnesty for those exiled or banned for crimes. Gerard had such decrees inscribed in the statutes of Parma.[11] He laid down specific norms for the return of those in exile, and established a routine procedure for removing bans imposed for lesser crimes.[12]

---

[8] '[Q]ui plenarie fecerunt, quidquid imperaverat servus Dei.'

[9] Rampona, 102: 'fé relasare tucti li presonieri delle carcere de Bologna'.

[10] Borselli, *Cron.*, 22.

[11] Parma Stat., 304. A practice that Daniela Gatti, 'Religiosità populare e movimento di Pace nel'Emilia del seclo XIII', in *Itinerari Storici: Il medioevo in Emilia* (Carpi, 1980), 98, considers an 'eccezionalità'! I examine Gerard's statutes in Ch. 8.

[12] Parma Stat., 313. His legislation in Modena (now lost) must have been remarkably effective since we are told that all exiles returned 'praeter quinque', *Ann. Vet. Mut.*, col. 60.

All those banned for any reason whatsoever could be absolved by the modest payment of 5s. parm.[13] The same statute protected them from any further punishment or condemnation for the same crime. He did make exceptions for some major crimes. First, this general amnesty did not apply to those banned for peace-breaking, clandestine murder, or perjury. In those sensitive and dangerous cases the routine, almost bureaucratic, procedure envisioned in the statutes could never apply.

Gerard established different norms for relaxing bans imposed before, during, and after his peace-making mandate. A procedure for relaxation of the ban by the *podestà* was stipulated for after the expiration of the mandate. For bans incurred before the commencement of Gerard's peace-making, the third Friday of July 1233, he allowed release if the one banned gave 'suitable security that he would keep the peace' or subscribed to a peace agreement negotiated by Gerard or his delegate. For those under the ban for debt or damage done to another—saving the case of feud (*guerra*)—a formal peace had to be arranged. That is to say, the person banned could be released at the consent of the one harmed, if he gave security to submit to judgement and discharge the penalty imposed. Gerard also announced that he would cancel all condemnations of those who had requested peace, made peace, or would make peace, before the feast of St Michael, 29 September 1233, the expiration date of his mandate for peace-making. This cancellation was good for any crime except that of peace-breaking committed after Gerard had received his commission as communal peace-maker.[14]

So although Gerard could relax the ban on peace-breakers condemned before his mandate, he did not have the authority to do so for those who broke the peace during or after it. He did, nevertheless, issue a decree to lighten the burden on such individuals. By one statute he gave those under the ban for peace-breaking from the feast of St Michael until the Assumption (15 August) of the following year licence to sell or dispose of their property within the city and district of Parma through a procurator.[15] This liberty removed any excuse for them to enter the city and allowed them to depart without excessive financial harm.

[13] Parma Stat., 312: 'pro quacomque causa, vel ex quacomque occasione'.

[14] Parma Stat., 312: 'Et hoc habeat locum in omnibus maleficiis commissis ante dictum diem veneris, nisi esset maleficium pro quo pax rupta sit postquam Frater Gerardus habuit potestatem faciendi paces.'

[15] Parma Stat., 314–15. The commune returned to a hard line against peace-breakers 6 years later when they declared a perpetual and irrevocable ban against clandestine murderers, perjurers, peace-breakers, and those who had committed murder within the Piazza Nuova (Platea Nova), Parma Stat., 313.

The preacher's functions as arbiter were not left to chance. They were clearly defined by the official acts of the contesting parties submitting their dispute to arbitration in accord with traditional legal forms. In northern and central Italy, the office of arbiter and the arbitration process had been well established by the 1230s. A comparison of reconciliation documents issued by John of Vicenza—the only friar for whom full sets of instruments remain—with the forms of mediation given in thirteenth-century Bolognese notarial manuals, such as those of Richardus Anglicus (1196), Rainerius Perusinus (1220s), Tancred of Bologna (1216, revised in 1234–5 by Bartholomew of Brescia), and Rolandinus Passagerii (later 1200s),[16] shows that the friars employed conventional notarial forms to give legal structure to their reconciliations.[17]

The procedure outlined in these books of notarial and Romano-canonical procedure is simple and clear, even elegant. The first step to a reconciliation was the drawing-up of an 'agreement to arbitration' (*compromissum*—the origin of the 'compromis' of modern international law) in which the parties submitted their dispute to arbitration by a third party. Then the arbiter prepared his judgement by hearing both sides of the case. Finally he issued his decision (*laudum*).[18] The parties could also terminate the process themselves by subscribing to a peace instrument (*instrumentum pacis et concordiae*). They could draw it up themselves or leave this to the arbiter. This was the 'normal' procedure.

We have several of the instruments that John of Vicenza produced. These include two agreements to arbitration, that between the bishop and commune of Bologna (19 April, 1233) and that among the cities of the Veneto (29 August, 1233). Four decisions by John survive: that between

---

[16] Rolandinus, in his *Summa Totius Artis Notariae*, i. On this manual see Enrico Besta, *Fonti, legislazione e scienza giuridica dalla caduta dell'Impero Romano al secolo decimosesto*, in Pasquale del Giudice (ed.), *Storia di diritto italiano*, i(2), (Florence, 1969), 829–30. This formulary, which later became the standard, dates from several decades after the Alleluia. It represents the traditions of older formularies and the development of practice.

[17] On the forms of arbitration and reconciliation, see: for instruments, Rich. Ang., 50–2; on judges, ibid. 104–14; on instruments of peace and concord, Rainerius, 54; on arbiters, Tancred (whose work treated both civil and ecclesiastical procedure), 103–8; for the most comprehensive treatment of instruments and peace-making, see Rolandinus, fos. 136–40. Rolandinus includes sample documents. On the notarial schools, see Gianfranco Orlandini, '"Studio" e scuola di notariato', *Atti del Convegno Internazionale di Studi Accursiani, Bologna, 21–26 ottobre 1963* (Milan, 1968), 73–95. For an example of the academic law behind the manuals, see Azzo dei Porci, *Summa* (Venice, 1596), 157 (on *arbiter*) and 173 (on *compositor amicibilis*).

[18] If the *arbiter* was following a strict legal format, the declaration of the *laudum* ended the process; it did not require formal acceptance by the parties. Tancred, 286, gives two forms of acceptance: 'sententia arbitri confirmata sit expressa partium volunte vel tacito decem dierum consensu'.

the bishop and commune (20 June, 1233), the Peace of Paquara (28 August, 1233), the order that the commune of Conegliano accept the jurisdiction of the bishop of Treviso (29 August, 1233), and the modification of that decree (30 September, 1233). In addition, there is the appeal of the commune of Treviso against the decision of 29 August. John's decisions carry the names 'instrumenta pacis' and 'concordiae', as well as 'lauda'.[19] In their completeness, these texts contrast sharply with the five brief, uninformative synopses preserved in the Bologna and Parma statute-books for other decisions by John and Gerard respectively.[20] Gerard's decisions are called simply 'peace agreements' (*paces*). Two of John's decisions are arbiter's decrees for Armanno of Porta Nuova; his third is the peace instrument of the parties involved.

After the parties involved in a dispute had decided, or had been convinced by others, to end their quarrel, two means of reconciliation were open to them. If they found themselves already in basic agreement, they could draw up an 'instrument of concord and peace' (*instrumentum concordiae et pacis*) and have it notarized. Otherwise they could formally enlist the aid of a third party. He could either help them reconcile their differences—act as a mediator—or issue his own settlement—act as an arbiter. The Alleluia preachers acted in both capacities. Since the second of these methods seems to have been more common during the peace-making of the Alleluia, it will be examined in detail. When a third party was to arbitrate a dispute, this was arranged in the *compromissum*.[21] As described in the notaries' manuals,[22] this instrument contained, in order, the following information: a description of the question under dispute, a statement by the parties that they had agreed to submit the question to some friendly person (*amicus noster*), the naming of that person, a description of his authority, the way in which he should render judgement,

---

[19] John's commission to arbitrate between the Bishop and Commune of Bologna is edited in Savioli, iii. 123–5; his decision in that conflict, in Savioli, iii. 126–33; his decision at Paquara, in *AIMA*, iv, cols. 1171–4. His instruments for the Veneto are edited in Verci, i. 103 (no. 70); i. 105 (no. 71); i. 108 (no. 75).

[20] The statutory remains of John's arbitrations are found in Bol. Stat., i. 449–50; those of Gerard in Parma Stat., 3, 216, 221, 292–3, 301, 302, 305, 307, 312.

[21] The origin of this procedure is found in Roman Law, C. 2. 55. 1. For an accessible synopsis of the classical forms, see W. W. Buckland, *A Textbook of Roman Law from Augustus to Justinian*, 2nd edn. (Cambridge, 1950), 527–33. For medieval jurists on civil and canon law arbitration, see Linda Fowler, 'Forms of Arbitration', in Stephen Kuttner (ed.), *Proceedings of the Fourth International Congress of Medieval Canon Law (Toronto, 21–25 August 1972)* (Vatican City, 1976), 133–47.

[22] For example, the tract 'De Compromissis' of Rolandinus, fo. 129ᵛ.

a promise to accept his decision, and, sometimes, the specification of a penalty for failure to do so.[23]

Both agreements to arbitration surviving from the activity of John of Vicenza follow this pattern. Rolandinus Passagerii, the noted thirteenth-century Bolognese jurist, tells us that when parties submitted their disputes to the judgement of a third party they could do so in two ways. One way was to submit a specific problem to the arbiter, who gave an answer to that question alone. In the second form the parties 'were able to submit to arbitration all the affairs and controversies of both parties generally, and in this form it is then called a "plenary agreement to arbitration"'.[24] An examination of John's two commissions to arbitrate suggests that his arbitrations followed both procedures. Following the normal form of an agreement to arbitration, John's 19 April, 1233 commission to arbitrate between the commune and bishop of Bologna first explained the nature of the conflict to be decided. It was 'over jurisdiction and exercise of jurisdiction in certain areas'.[25] It was a commission of the more restricted type. It would seem likely that a good number of disputes mediated by the friars, such as John's judgement on Armanno's garden, or Gerard's for the hospital, were also of this type. Sometimes, however, the controversies were so complex that the parties handed them over to the mediator, who received a broad commission to work them out. John's 29 August, 1233 commission is an example of this procedure. It tells us that the parties 'choose, without fear, as arbiter, mediator and friendly reconciler, Brother John of Vicenza of the Order of Preachers here present, for all questions, disputes, controversies, injuries and feuds that have heretofore existed among them'.[26] Here the parties

---

[23] A penalty was provided since the agreement to arbitration was not, of itself, legally enforceable. See Tancred, 103; or Azzo, *Summa*, 2. 3. Even as a *pactus nudus*, it was, however, considered enforceable in Canon Law, X 1. 35. 3 and was considered to give rise to an action: C. 12 q. 2 c. 66 and C. 12 q. 5 c. 5. The argument of the canonists, as in C. 22 q. 5 c. 12, was that violation of one's word and violation of an oath were both sins. John invoked ecclesiastical sanctions against those who violated his decrees.

[24] Rolandinus, fo. 129ᵛ: 'Item compromitti posset generaliter de omnibus rebus et controversiis utriusque partis, et dicitur tunc plenum compromissum isto modo.'

[25] Savioli, iii. 123: 'super iurisdictionibus et occasione iurisdictionum quarumdam terrarum, sc. S. Johannis in Persiceta Unzoli dulioli Castri Episcopi Podii de Maxumatio Maxumatici Ulgiani Flexi, sc. infra locum qui dicitur infra bannum d. Episcopi Montiscavallorii et Arzele a parte inferiore se ubi habitant homines super proprietate Episcopatus Bon.'

[26] Verci, i. 103: 'Compromiserunt se sine omni timore in fratrem Johannem Vicentinum de ordine fratrum Predicatorum presentem tanquam in arbitrum, et arbitratorem, et amicabilem compositorem de omnibus questionibus, litis, controversiis, injuriis, et guerris, que hactenus fuerunt inter eos et ipsorum comunia.'

left it to the arbiter to work out the nature and importance of the issues involved and develop a comprehensive agreement to end hostilities.

John's commissions to arbitrate did not simply bestow power. As the notarial forms would imply, they imposed certain restrictions on the negotiations. In the Bologna commission John spoke as if it was he himself who laid down the norms. Speaking in the first person ('descerno', 'volo', 'praecipio', etc.), he promised, in particular, to show no favoritism, and to exact an oath from the parties not to harm their adversaries and to allow them freedom of movement.[27] Furthermore, he promised to exact an oath that all property occupied or carried off during the conflict be returned and that all condemnations and bans imposed during the conflict be rescinded. Finally he excommunicated, and called down the wrath of God and of the Apostles Peter and Paul upon any member of either party who disturbed the peace process, or refused to accept him as mediator.[28]

This document shows a remarkable similarity to standard notarial forms. John clears away impediments to reconciliation in much the same way the commune of Parma did when it commissioned Gerard. In both cases means were specified for reducing fear and tension. The principal difference is that the commune of Parma, by its own statutes, guaranteed freedom of movement and forbade continued hostilities—before any parties had drawn up an agreement to arbitration. In Bologna, one disputant was the commune itself, so such preliminary measures had to be provided for in the agreement to arbitration itself. It is reasonable to assume that even in Parma, where statutes gave norms for entering mediation, the mediator and the parties involved used their agreements to arbitration to reaffirm communal norms or establish others according to their particular situation.

Part of the agreement to arbitration defined the specific role to be played by the mediator himself. John's documents follow conventional notarial form and name him, in the Bologna arbitration, as 'mediator and friendly reconciler' (*arbitratorem et amicabilem compositorem*) and, in the Veneto arbitration, as 'arbiter,[29] mediator, and friendly reconciler' (*arbitrum et*

---

[27] *AIMA*, iv. 1172: 'Item praecipio sub poena Sacramenti, quod illi qui steterunt in Civitate et in Villis nullam injuriam inferant illis, qui revertentur, sed tamquam fratres eos benigne ac pacifice recipiant, atque tractent.'

[28] *AIMA*, iv, 1173–4.

[29] There is a recent full-length study of the office of *arbiter* in the middle ages: Luciano Martone, Arbiter-Arbitrator: *Forme di giustizia privata nell'eta dell diritto commune* (Naples, 1984). But he treats notarial and legal documents themselves, not the way these offices functioned in specific cases. A technical treatment can be found in Karl S. Bader, 'Arbiter,

arbitratorem et amicabilem compositorem).[30] These documents appointing John to arbitrate call themselves *compromissa*.[31] These terms themselves suggest a great deal about the functions performed by the friars during their mediation. In the original Roman Law form of the agreement to arbitration, the third party appointed to settle the dispute had been called simply an arbiter (*arbiter*). In the classical period of Roman Law, an arbiter had been someone with special expertise in the matter to be adjudicated. Reconciliation by arbiter was common in disputes where there was a need for more-than-average discretion.[32]

The notary Rolandinus helps us to see how legists understood this functionary in the thirteenth century.[33] He tells us that in every particular the function of an arbiter (*arbiter*) is identical to that of a judge (*iudex*).[34] That is to say, after his selection, the parties presented their arguments to him in the form of written briefs (*libelli*), at a time and place specified in the agreement to arbitration. The process ended with the arbiter's declaration of his judgement (*sententia*). Since this judgement could not give rise to another action, he enforced it by providing some sanction for failure to comply. No appeal was possible, and the arbiter had to follow strict legal procedure.

John's commission to arbitrate in Bologna names him not 'arbiter' (*arbiter*) but 'mediator' (*arbitrator*), a word synonymous in notarial usage with 'friendly reconciler' (*compositor amicabilis*).[35] Neither of these titles

---

Arbitrator seu Amicabilis Compositor', *Zeitschrift der Savigny-Stiftung für Rechtsgeschichte* (*Kan. Abt.*) 77 (1960), 239–76.

[30] This longer form also appears in John's decree for the bishop and commune, Savioli, iii. 129; in his 29 August instrument at Paquara, *AIMA*, iv. 1170; and in all the notaries' formularies. The formula also appears in arbitrations by seculars; for example, in that of Otto of Mandello and Rainier Zeno between Treviso and Padua promulgated on 11 Sept. 1235, Verci, i. 118, where they are called, 'arbitri seu arbitratores seu amicabiles compositores electi'.

[31] For example, that in Bologna, *AIMA*, iv. 1170.

[32] Buckland, *A Textbook of Roman Law*, 529, who draws on Cicero's description of the office.

[33] Rolandinus, fo. 130^{r-v}: 'Quid sit arbitrium'.

[34] On procedure before a *iudex*, see Rich. Ang., 104–11, and, for a more elaborate version, Tancred, 196–201. These two authors treat action before a magistrate, not an *arbiter*. The major part of Richardus' treatise covers the various forms of *libelli*. Those of disputants before an *arbiter* would have been identical. Rainerius, 190–218 describes similar *libelli*. Unfortunately, no *libelli* remain from the 1233 arbitrations for comparison with those in the manuals.

[35] Charles Du Cange, *Glossarium Mediae et Infimae Latinitatis*, ii (Paris, 1937), 470: 'Arbitrator, id est amicabilis compositor'. The 'friendly' here refers to the form of the reconciliation, not to any personal quality of the reconciler.

appears in the corpus of Roman Law.[36] Rolandinus carefully describes this office:

A mediator is that one to whom some parties entrust a controversy for mediation as they would to a good friendly man so that their dispute can be settled by his counsel and authority. Such mediators are not bound to follow the procedures given for legal judgements, which the arbiter must carefully follow.[37]

The weakness of such a moderator was that one or both parties could simply refuse to obey his decision, if they thought it unjust. In contrast, the judgement of an arbiter (*arbiter*) was to stand and could not be challenged unless it was expressly contrary to civil or canon law.[38]

Rolandinus also tells us that, by a legal sleight of hand, thirteenth-century disputants could get the best of all possible judges. They could appoint the umpire of their quarrel simultaneously both 'arbiter' and 'mediator'. In that way he could have the freedom of procedure enjoyed by a mediator and the power of definitive judgement proper to an arbiter. This is exactly the kind of commission the preachers received. The one apparent exception is John's commission in Bologna, where he was named simply 'mediator'. Later, when he rendered his decision, John called himself by both titles. Perhaps at the beginning the parties were unwilling to grant him the full authority of an arbiter and so named him only as mediator. Whatever the case, he had full authority by the time he rendered his decision.

An example of an appeal from one of John's judgements has also come down to us. It concerns the dispute over which his Veneto commission had empowered him. In this appeal, the commune of Conegliano, dissatisfied with John's decision, does not simply ignore it, as they might have, had it been the decision of a mere friendly reconciler. Instead they denounce it as against the law and appeal to higher courts, that is, to the emperor and the pope. They condemn John's decision in the strongest

---

[36] 'compositor' does appear, but as the title of the author of laws, e.g. 'compositor juris enucleati', D. 1. 4. 3; C. 6. 28. The closest thing to an *amicabilis compositor* in Roman Law is found in Nov. 9. 10., where a bishop acts as a mediator.

[37] Rolandinus, fo. 134ʳ: 'Arbitrator autem est ille in quem alique partes quandoque compromittunt tanquam in amicum bonum virum ut eius consilio et authoritate aliqua eorum discordia decidatur, et isti arbitratores non astringuntur ad servandum ordinem datum in iudiciis, quem tamen arbiter bene servare teneretur.'

[38] Rolandinus, fo. 134ʳ: 'Et quod dixi sententiae arbitri standum esse sive aequa sive iniqua sit, intellige verum esse, dummodo non esset expressim contra leges vel sacras constitutiones.'

language possible.[39] This is in accord with the circumstances under which the decision of an arbiter could be rejected.

The extant peace-instruments explain what the parties expected of their 'friendly reconciler'. The extent of John's authority is clearly specified. He received 'full and absolute power over the aforesaid controversies' (*plenam et absolutam potestatem super predictis controversiis*). This gave him the power of 'decreeing, ordering, and ruling whatever he himself wanted' (*laudandi precipiendi statuendi quidquid ipse voluerit*) over this and any future disputes between the two parties. He also received authority to modify or re-interpret his decisions however he might want.[40] John did just that when he quashed his initial 28 April decision for the bishop and commune of Bologna and replaced it with the one of 20 June that has come down to us.[41] Awesome as this description of his power sounds, there is nothing extraordinary about it. The language is, in fact, identical with that used to describe the authority of an arbiter-mediator in Rolandinus.[42]

INSTRUMENTS OF RECONCILIATION

Once a mediator had been chosen and commissioned, the dispute came into arbitration or negotiation. Since the friars had the freedom of action proper to a mediator, they were free to follow whatever procedure they wished. The Theobald story shows one form of procedure, the less formal, in action. After an aborted reconciliation, Theobald ordered the parties to dine together, talk it over, and return the following day, when he would draw up the peace instrument. The narrator tells us that they complied

---

[39] Text in Verci, i. 107. Their representative writes 'Dico Sententiam, Arbitrium, Preceptum, et diffinitionem nullam, et iniquam, et contra jus. Et si aliqua esset in scriptis ad Dominium Papam, et ad Dominum Imperatorem pro eis appello, et supplico, et Apostolos instanter peto, et me pro eis in protectione domini Pape pono.'

[40] Savioli, iii. 123: '[C]ompromiserunt [ambe partes] . . . in d. Fr. Johannem tamquam in arbitratorem et amicabilem compositorem concedentes eidem plenam et absolutam potestatem super predictis controversiis et aliis que in futuram apparerent inter ipsas laudandi precipiendi statuendi quidquid ipse voluerit sive coniunctim sive separatim de questionibus iam ortis vel que in futurum orirentur vellet precipere seu statuere seu arbitrari dantes et concedentes eidem d. Fr. Johanni liberam potestatem si quid aliquando preciperit quod videretur vel diceretur obscurum vel dubium quod ipsa possit illud declarare et interpretari et determinare quandocumque et qualitercumque voluerit et si quid etiam statuerit quod in futurum videretur ei mutandum quod ipse possit illud mutare sicut sibi videbitur.'

[41] See Sutter, 92.   [42] Rolandinus, fo. 129ᵛ.

and then 'completely fulfilled all that the Servant of God had ordered'.[43] Dinner is hardly a formal legal setting, so one can imagine a certain give-and-take in the negotiation process. If Theobald came to dinner, he himself could have moderated the discussion. The instrument drawn up by him would be the result of bargaining among the parties and thus more or less the work of the disputants themselves. If this was the case, Theobald did not even have to render a decision in the case; instead, he could simply help draft the *instrumentum pacis et concordiae*.[44]

The friars also acted as true arbiters, issuing a formal judgement after reviewing the case. John quashed and revised his first decision in the reconciliation of the bishop and commune of Bologna. This need for revision suggests that the initial decree was not completely satisfactory at the time of its publication. So it seems likely that John and other peace-makers sometimes gave the contestants only a general idea of their intentions before issuing their decision. Thus he need not have submitted trial proposals to the contestants before he published his actual decision. Here, if the contestants failed to accept it, the arbiter's only recourse was to issue a new and revised decree—or give up and go home in frustration, perhaps invoking divine intervention. When genuine arbitration occurred, the arbiter issued what the notaries called an *instrumentum arbitrii et laudi*.[45] Such an instrument contained the following information: the name of the parties and the arbiter, a description of the dispute, some reference to the means used to investigate and settle it, the decision of the arbiter, and an indication that the parties were present to accept it. John's decision for the bishop and commune followed this order precisely.

The language used in other documents suggests that many recon-ciliations worked by the friars were not formal arbitrations. The most commonly found name for reconciliation documents is 'peace instrument' (*instrumentum pacis*), a variation of which, 'instrument of harmony and agreement' (*instrumentum concordiae et compositionis*), appears in one Parma statute.[46] In theory, this document gave final form to what the parties

---

[43] *VF*, 225: 'qui plenarie fecerunt, quidquid imperaverat servus Dei'.

[44] Rolandinus, fo. 139[r].

[45] Rolandinus, fos. 138[v]–139[v], gives the form of these instruments.

[46] Parma Stat., 197. The word *compositio* here reflects John's description as *amicabilis compositor*, Savioli, iii. 123. Such instruments are not unique to northern Italy. See William Bowsky, 'The Medieval Commune and Internal Violence: Police Power and Public Safety in Siena, 1287–1355', *American Historical Review*, 73 (1967), 123, on their use in Siena. Tuscan texts are collected in *Collectio Chartarum Pacis Privatae Medii Aevi ad Regionem Tusciae Pertinentium*, ed. Gino Masi (Orbis Romanus, 16; Milan, 1943). I know of no comparable collection for the north.

had agreed among themselves.[47] Sometimes the friars simply helped put the reconciliation in writing and did not formally arbitrate at all. The greater the number of persons involved, the more likely the forms employed resembled formal arbitration, while the more restricted and private the dispute, the more they resembled informal 'friendly' reconciliation.[48]

Occasionally reconciliation documents carry names such as *pax*, *concordia*, or *treuga*.[49] To some extent, more precise meanings can be given to these seemingly synonymous words. All put an end to disputes. The broadest is *pax*, 'peace', which can be used for all reconciliations.[50] There are two types of peace, private and public, the former between persons and the latter between cities. This latter is also called *treuga* or *treugua*, 'truce'.[51] Rolandinus tells us that in his time the statutes of the Italian cities regularly enforced *paces*. He restricts the use of the word *concordia*, 'agreement' to those more private reconciliations that the communes did not enforce.[52] Thus, there was a public element in a *pax* that was lacking in a *concordia*.[53]

The Theobald story gives us a hint of how a reconciliation acquired public status and enforceability. This story highlights two distinctively public acts. The first-mentioned, though in some ways less important, is the kiss of peace—that gesture that the murdered man's brother, in a fit of anger, refused to give the murderer. The model for a peace instrument given by

---

[47] Rolandinus, fo. 139^r, describes the legal status of such a document and how it was drafted.

[48] Professor James Gordley of the Law School of the University of California at Berkeley tells me that the same pattern is seen in modern law: corporation law being formal and impersonal, family law being more flexible and adapted to the personalities of the parties.

[49] For examples, *pax*: Vic. Stat., 233, 245, Parma Stat., 198, 199; *paces, concordiae, et sedationes*: Vic. Stat., 15; *treuga*: AIMA, iv. 1172, Vic. Stat., 73, 117; *pax vel treuga*: Vercelli Stat., col. 1128, 1129.

[50] Rolandinus, fo. 139^r, gives the following definition: 'Pax est discordiae finis, unde quotiescumque aliqui guerram ad invicem habentes volunt ab omni guerra et discordia, et ab omni offensione perpetuo discedere, dicit tabellio, tales fecerunt invicem pacem.'

[51] Rolandinus, fol. 139^r, implies that *treugua* is less permanent than *pax*, which is, by definition, perpetual: 'Treugua est conventio seu confidentia quedam facta in longum tempus, id est de non se lacescendo, id est de non provocando se ad bellum, vel diffinitur sic.'

[52] Ibid., fo. 139^v. Tancred, 250, also describes 'private instruments' which do not have 'publicam auctoritatem', but these do not seem to include *concordiae*.

[53] These distinctions cannot be pressed too far. Rainerius, 54, has the same notarial form for a *pax*, a *concordia*, and a *treugua*. The language of extant instruments themselves is much more fluid than that of Rolandinus. John of Vicenza refers to his decision in the mediation in Bologna as a 'decree of peace and harmony' (*laudum pacis et concordiae*), a formula which blends names of instruments distinct for Rolandinus.

Rolandinus specifically mentions this kiss of peace as part of the form.[54] The second act to appear in the story, although it preceded the kiss in the ceremonies of reconciliation, was the drafting and publication of the peace instrument itself. Statutory evidence sheds considerable light on both the publication of the instrument and the kiss. John himself describes the manner in which the instruments were published, at least in the more important reconciliations, in the preamble to his Bologna decree of 20 June 1233:

I, brother John, now of Bologna, but formerly by birth of Vicenza, of the Order of Preachers, arbiter, mediator and friendly reconciler, chosen on the one part by Lord Henry, by the grace of God bishop of Bologna, in his own name and that of his bishopric, and on the other part by the commune of Bologna, in the Special and General Councils, assembled by the sounding of bells according to the custom of the city of Bologna, and in which council is also convened the directors of the Arts and the Arms and many others, just as was specified in the agreement for arbitration . . . for the sake of peace and concord, declare, order, and decide . . .[55]

So the more public form of peace instrument received a public reading in the presence of the parties, other witnesses, and a sizeable crowd. Such a forum put public pressure on the parties to accept the arbiter's decision and keep it in the future. In John's case the document became a legally

---

[54] Rolandinus, fo. 139[r]: '[F]ecerunt invicem osculo pacis vicissim inter eos veniente, pacem perpetuam.' On the legal role of the kiss in Roman Law, see Mary Brown Pharr, 'The Kiss in Roman Law', *Classical Journal*, 42 (1946/47), 393–7. On the use of the kiss in liturgical ceremonies of reconciliation and peace, see the cursory study of Nicolas James Perella, *The Kiss Sacred and Profane: An Interpretive History of Kiss Symbolism and Related Religio-Erotic Themes* (Berkeley, Calif., 1969), 12–25. The kiss was forbidden to heretics; see Elizabeth Vodola, *Excommunication in the Middle Ages* (Berkeley, 1986), 52–3.

[55] Savioli, iii. 126: 'Ego Frater Johannes de Bononia nunc qui olim fui de Vicentia oriundus de Ordine Fratrum Predicatorum arbiter et arbitrator seu amicabilis compositor electus a domno Henrico dei gratia Bon. Episcopo nomine suo et sui episcopatus ex una parte et a Comm. Bon. ex altera in consilio euisdem Civitatis tam speciali quam generali coadunato ad sonum campanarum sicut consuetudo est Civitatis Bon. in quo etiam consilio convenerunt ministrales artium et armarum, et alii plures sicut continetur in compromisso . . . pro bono pacis et concordie laudo, precipio et arbitror . . .'

Professor Gerard Caspary of the University of California, Berkeley, tells me that early 13th-cent. civilians such as Azzo recognized the sounding of a bell to signal a meeting of a municipal assembly as the sign that an official act was to be performed. In Bologna the sounding of a bell appears about 1217, at the time when the populars received a role in the government through the creation of the Consilium Generale. The older Consilium Credentie, for which it seems bells were not sounded, continued to function for some time. Since, unlike the older body, the newer one's meetings were announced by a bell and criers, it became known as the 'Consilium ad sonum campane congregatum et per precones clamatum'. See Alfred Hessel, *Storia della città di Bologna dal 1116 al 1280*, ed. and trans. Gina Fasoli (Bologna, 1975), 173–4.

public act also because the same notary who signed the agreement for arbitration, Bolognitto of Strada Maggiore, himself notarized it.

Although the production of this agreement may have been typical of Bologna only, it is comparable to the series of Parma statutes issued to regularize and make enforceable the peace-making of friar Gerard. These give us some idea of the variations possible in less weighty matters. It was by no means necessary that those making peace appear before the friar in person.[56] Gerard appointed vicars who acted in his place, a procedure also followed by John of Vicenza.[57] Conversely, peace could be concluded not only by the parties involved but also, for a minor, through a guardian (*curator*), and, for an adult, by a proctor (*procurator*).[58] Gerard took the trouble to issue a statute that would protect the inviolability of accords made by these agents, if at some later date an agent denied that he had functioned as such. He declared the written instrument itself sufficient proof of agency by the agent named.[59] Clearly, someone had once tried to escape the arbiter's decision or the perpetuity of the peace by pleading the invalidity of the instruments contracted through an agent. Or some may have attempted, in spite of a formal peace, to take advantage of minors. This was a danger that the friars attempted to remedy by their statutory legislation.

After the peace instrument, the most visible public sign of reconciliation was the kiss of peace. A modern reader might be inclined to think that the phrase 'giving the peace' simply shows that a peace had been contracted. Rather, the phrase refers to the public giving of a ritual kiss, most likely on the lips. The act had a sacramental flavour and called to mind the giving of the peace before Holy Communion at Mass. The kiss was not absolutely necessary for contracting peace, as a text in the statutes of Verona from 1228 shows: 'If anyone breaks a peace, whether sealed with a kiss or without a kiss, with any person under the ban, he is held to make amends in the same manner, that is by money.'[60] So, by the 1220s,

---

[56] As we know from Tancred, 108.

[57] Parma Stat., 301, gives equal authority to *paces* whether mediated by Gerard or by his vicar. John also acted through nuncios and vicars, as we know from the 3 Aug. 1233 instrument (in Verci, i. 102), issued by his nuncios Gavardo and Berardo, in which he ordered the commune of Conegliano to release some captured citizens of Treviso.

[58] Rolandinus, fos. 139ᵛ–140ʳ, explains the use of such agents.

[59] Parma Stat., 302. These offices of *curator*, *tutor*, and *procurator* seem identical with those functionaries in Roman Law.

[60] Ver. 1228 Stat., 68: 'Si quis pacem fregerit osculo, vel sine osculo firmatam alicui bannito, eodem modo teneatur emendere, scilicet pecunialiter, ac si esset in treva.' The text is repeated verbatim in Ver. 1276 Stat., 415.

in Verona at least, a distinction could be made between those accords
sealed with a kiss and those not, even if sometimes, as in this case, no
distinction was made between the punishments for breaking one kind or
the other.

A Bologna statute of 1252, specifying the point at which the peace
becomes binding, says this: 'We then understand there to be a peace after
the kiss has been given by the principal persons and his [that is, the
banned individual's] agent.'[61] Here, when a party exiled for violence
made peace with the victim and his family through an agent, the peace
took effect from the exchange of their kiss. Accords contracted directly by
the parties also commenced with the giving of the kiss. The 'giving of the
peace' was a dramatic gesture and by nature more striking than the
notarization of the instrument. It signalled the conclusion of the reconciliation, not its beginning. So, when Theobald called the murderer and
the murdered man's brother out to give the kiss, it is certain that
negotiation, and perhaps even a written instrument, had preceded that
dramatic event. The existence of accords contracted without the kiss, as
implied by the Verona statute, must be allowed, but these would, of
their nature, have been less public or solemn. The kiss in public: that was
the definitive, legal, end of the feud.

The public reconciliation described in the Theobald story—a public
giving of the peace—would immediately have suggested itself to the
friars as the suitable formality for sealing their accords. It also provided a
short cut for giving reconciliations public status, and could make for a
very impressive and memorable show. From the legislation recognizing
the right of the friars' vicars and other agents to arbitrate disputes, it is
clear that many controversies might require reconciliation during the
course of a single peace-campaign. In practice, this almost routine administration of justice far outweighed in bulk the few dramatic reconciliations
of bitter enemies for which narrative descriptions or peace instruments
survive. To seal these 'routine' reconciliations, kisses were exchanged,
many at a time when the friars 'reconciled' groups of disputants publicly,
probably after a sermon on peace. This formality could be saved until the
climax of the campaign, when it would have made the greatest impact
on observers. It is no wonder that the chroniclers speak of mass reconciliations during which many enemies kissed for the first time.

---

[61] Bol. Stat., i. 267: 'Pacem autem intelligimus ab oscullo interveniente per principales
personas et procuratorem eius.' This mirrors closely the wording about the kiss in
Rolandinus on an *instrumentum pacis*.

Did the friars also formalize peace in some less public forum, drafting what Rolandinus called *concordiae*, in which simple notarization of the instrument sufficed for validity? Without any evidence for or against this, one cannot tell. It would seem from the 1228 Verona statute, which mentions accords made 'without a kiss', that the kiss was not absolutely necessary for contracting a peace. But the dramatic potential of mass giving of the kiss would have influenced the friars to steer as many reconciliations as possible, even minor ones, into the public forum.

A second look at the reconciliation brought about by friar Theobald is in order to provide context for an analysis of the sanctions employed to prevent violation of a newly contracted peace. The story gives us a good idea of external appearances of a public reconciliation. But appearances are deceptive. The narrator would have us think that Theobald simply noticed the presence of the two feuding parties and spontaneously called them out to make peace. That this event took place 'on a day when Theobald had reconciled a large number of people' should put us on our guard against such a conclusion. Considering what we know about peace-making, Theobald certainly orchestrated these reconciliations after or during a solemn sermon on the theme of peace. Since the number reconciled that day was reportedly large, the events probably marked the high point or the end of a pacification campaign. So, those present, including the murderer and his victim's brother, had already worked out their differences. They were present for the final public sealing by a kiss of their already-reached agreements, or to receive publicly Theobald's prepared 'decrees of concord'. Theobald's abrupt command that the brother and murderer come forward and exchange the kiss of peace confirms this. A summons to kiss was nothing unusual: the kiss did not of itself create a reconciliation but placed the final seal on a completed peace-agreement.

The reconciliation in this story was far less 'spontaneous' than the author implies. It would have been the result of some more or less protracted process of mediation or arbitration. Theobald was a master of his art. His vigorous invocation of God and his tactful miracle marvellously smoothed over a final flare-up of resistance to something already arranged. If there is a doubt about this, one need only contrast Theobald's treatment of the first brother with his conduct towards the second threat to reconciliation—the victim's other brother. Here he commands, not another immediate kiss of reconciliation, but a 'negotiation over dinner' followed by the drafting of a formal peace-instrument (*instrumentum pacis*).

This implies that the agreement already reached had not included the second brother. Perhaps his arrival even took Theobald by surprise! The second brother had to undergo the normal series of steps to a reconciliation: mediation or arbitration, composition and acceptance of a peace instrument, and formal public sealing of that agreement by a kiss. The busy peace-maker Theobald could then add another dramatic reconciliation to his successes of the day.

The elaborate legal machinery of mediation and reconciliation, even if it was the most important part of peace-making, lacked the dramatic excitement of the publication of the arbiter's decree and the parties' kiss of peace. The legalities quickly disappeared from the memories of observers. What struck their imagination was peace-making's dramatic concluding rituals. Keeping in mind that the hard bargaining was invisible to the crowd, the people's view of the friars' peace-making as an awe-inspiring display of divine power is not surprising. Its accomplishment was the crowning miracle in the repertoire of the wonder-working Alleluia preachers. Peace and reconciliation—occasionally, if only rarely, worked out on the spot—was the most valuable and coveted product of the Great Devotion.

After the reconciliations and conversions of peace-making had ended, more mundane business remained. The peace had been sealed, either by document, by kiss, or both, and now some means had to be found to prevent its violation. Legally, in the eyes of the jurists at least, the decision of a mere reconciler (*arbitrator* or *amicabilis compositor*) was not in itself legally binding.[62] The notarial manuals and statutes give various indirect means to guarantee observance. Rolandinus devotes over half of his treatment of the *instrumentum concordiae et pacis* to marriage contracts.[63] Betrothal of the children of reconciled parties was a possible way to ensure peace between reconciled families. At the Peace of Paquara, John of Vicenza used this very method. He ordered the marriage of the son of Azzo d'Este and the daughter of Alberico da Romano, thus uniting the leading families of the two major warring parties in the Veneto.[64] If

---

[62] Tancred, 103: 'Et in eos sub poenae stipulatione compromittitur, ut metu poenae ipsorum sententiae stetur; quoniam sententia arbitri non valet sine paenae promissione vel pignorum datione.' He then quotes C. 2. 56. 1. It would appear that the communes did not automatically enforce *paces* until much later; in Florence, for example, not until the mid-14th cent.; see Kuehn, 'Arbitration and Law', 295–7.

[63] Rolandinus, fos. 139ᵛ–140ʳ.

[64] *Liber Regiminum Padue*, ed. Antonio Bonardi (RIS² 8:1, Città di Castello, 1905), 310; Maurisio, 32, records the reaction to this decision: 'dictum ab omnibus cum magnis

preachers used this tactic in other Alleluia reconciliations, there is no evidence to that effect. Another way to ensure observance was to take a pledge from the parties. Gerard bound the *podestà* of Parma to exact suitable pledges (*ydoneas securitates*) from those wanting to make peace.[65] Anthony of Padua also took in pledge (*pignori*) the property of those who had violently seized the goods of others.[66] Undoubtedly, were the documentation more complete, it would contain examples of this practice during the Devotion.[67]

The simplest, most direct way to ensure the observation of the mediator's decree was for him to include sanctions, usually a fine, in the peace instrument itself.[68] In the peace decree for Armanno recorded in the Bologna statute-book, John did just this when he imposed a fine of £1,000 bol. for violation of the decision.[69] A fine was all well and good, but it still required someone to exact it. In Armanno's case, surely, one reason—perhaps the principal reason—that John entered the instrument in the Bologna statutes was to bind the *podestà* to enforce it with his ban and exact a fine from those banned for violation. The need for such enforcement by the *podestà* explains why peace instruments, which would appear to be matters between private individuals, were routinely entered in municipal lawbooks. In comparison with John's decree at Bologna, the other reconciliations registered in the statute books are mere abstracts. Most commonly, the statutes simply record that an agreement had been reached.[70] To provide for enforcement by the ban, a simple provision in the statutes was apparently adequate.

laudibus fuit gratissime confirmatum'. This agreement appears to have been a betrothal. On the difficulty of enforcing betrothals and the 'suspended' marriages of minors over the age of seven, see R. H. Helmholz, *Marriage Litigation in Medieval England* (Cambridge, 1974), 98–100. Only after an actual marriage between males over 14 and females over 12 could the parties be certain that the contract would hold.

[65] Parma Stat., 304.

[66] *Assidua*, 343.

[67] One might wonder if, along with pledges, the friars required anyone to stand as security (*fideiussor*) for the reconciled parties. This was a Florentine practice as in the model *instrumentum treuguae* of *Formularium Florentinum Artis Notariae (1220–1242)*, ed. Gino Masi (Orbis Romanus, 17; Milan, 1943), 44–5: 'Insuper talis pro tali fideiubens et cetera et talis pro tali et cetera'. I find no evidence of this practice in extant Alleluia instruments or in the Bolognese notarial manuals.

[68] Use of force to compel observance remained rare until the 1300s. On later peace-instruments and use of police authority to enforce peace pacts in Tuscany, see Bowsky, 'The Medieval Italian Commune', 1–17.

[69] Bol. Stat., i. 449.

[70] For example, the brief notice of a peace pact between Parma and the Hospitalers, Parma Stat., 198; or that between the city and *ministri* of a bridge on Strada Claudia, in Parma Stat., 199.

Another means of providing communal enforcement of peace agreements, less cumbersome than the individual registration of arbiters' decisions, was the declaration of a statutory punishment for all peace-breaking. Strictly speaking, the heinous crime of 'peace-breaking' was not any ordinary act of violence but only one that violated a formally ratified peace-agreement. In Parma, a 1233 chapter of the municipal statutes protected this principle against abuse:

The *podestà* or his judges or anyone else cannot and ought not place anyone under the ban for peace-breaking . . . unless it is clear to him, from public written peace instruments or worthy witnesses, that the one who is accused had made a peace agreement and broken it.[71]

When a statutory definition of peace-breaking existed, the friars could draft sanctions against such an act, or perfect those punishments that already existed. By doing this they guaranteed that their reconciliations would have lasting effect, even if they had not enrolled the individual reconciliations, one by one, among the laws of the city.

The anxiety that general laws against peace-breaking were not in themselves adequate led to legal recognition of the friars' reconciliations as a whole. This was done, in the case of Gerard, by placing the reconciliation commission and acts in the statute book itself.[72] The peace accords that he negotiated were to have the force of law. Gerard bound the *podestà* and his judges under oath to preserve inviolate, and cause to be observed, all decrees given by himself or his representatives during the peace-making process.[73] The means to this end was the imposition of the ban, a punishment that could be imposed on culprits individually or, as Parma eventually did in 1239, through the public proclamation of a list of those who had incurred the punishment for peace-breaking.[74] Other communes, including those not visited by the Alleluia preachers, decreed similar enforcement of peace instruments.[75]

---

[71] Parma Stat., 309: 'Capitulum quod Potestas, vel sui judices vel aliquis alius, non possit nec debeat aliquem ponere in banno pro pace rupta [Additum est: 'et de morte furtiva'], nisi ei liquidum fuerit instrumentis publicis vel testibus ydoneis quod ille, qui accusatus fuerit, pacem fecisset et rupisset.'

[72] Examples: for John of Vicenza, Bol. Stat., i, 448–9; for Gerard, Parma Stat., 194, 198–9.

[73] Parma Stat., 302: 'Item statuit et firmavit quod Potestas et judices ejus et Consules Communis et omnes officiales teneantur expressim ex debito juramenti facere servari et attendi et inviolabiliter omnia judicia, sentencias, pronunciationes et lauda quae ipse Frater Gerardus per se vel per suos vicarios et nuncios speciales super pacibus factis et faciendis faciet et pronunciavit.'

[74] Parma Stat., 313.

[75] Vic. Stat., 130: orders the *podestà* to compel observance of peace; Ver. 1228 Stat. (ch.

Beyond temporal penalties there remained one other means of enforcing the agreements—recourse to divine punishment. This was a favoured tactic of the revivalists of 1233. While the curse, or even excommunication, imposed by the mediator may have been used to enforce decisions for private individuals, the most visible examples of its use are in John's arbitrations at the supra-communal level. At Paquara, when John declared the terms of the peace between the cities and factions of the Veneto, he 'cursed the crops, vines, trees, livestock and all the goods' of those who dared to violate his ruling.[76] This blanket curse must have occurred during John's sermon on peace, since in his formal instrument he restricted himself to merely 'cursing, excommunicating and anathematizing, by the authority of God the Father, Son, and Holy Spirit, and of the Apostles Peter and Paul' those who rejected or violated his decree.[77] In his sentence in the dispute between the commune and bishop of Bologna, John invoked both temporal and spiritual sanctions:

I command and order that each and every part of the above decree be observed inviolate forever, by the authority of the Father, Son, and Holy Spirit, by the power of Our Lord Jesus Christ, and under the penalty of two thousand marks of silver.[78]

The exact nature of the divine vengeance against those who broke the peace he left unspecified. Perhaps the presence of a secular authority in Bologna to enforce the temporal penalty rendered further elaboration of supernatural punishments unnecessary.

More effective for preserving the peace than any sanction was the removal of the causes from which disputes arose. Ultimately peace would only prevail if the grievances that triggered war had ceased. Although their complete removal lay beyond the powers of any friar-arbiter, the cities did give the friars the opportunity to remedy some of the causes of unrest

---

86), 67–8: imposes the ban on all who break the peace; Vercelli Stat. (§83), col. 1128–9: exiles peace-breakers unless they make peace with their victim or his heirs, and imposes fines according to the peace-breaker's wealth.

[76] Rolandino, 45: 'et rebellium maledixit fruges, vineas, arbores et bestias et omnia que haberent'.

[77] *AIMA*, iv. 1173: 'Ex parte igitur Dei Patris, et Filii, et Spiritus Sancti, autoritate quoque Apostolorum Petri et Pauli maledico, excommunico, anathematizo illam partem, vel illos, sive illum de alterutra partium, qui hanc pacem non susceperint.'

[78] Savioli, iii. 132: 'Et hec omnia et singula superius dicta ab utraque parte autoritate Patris et filii et Spiritus Sancti in virtute nostri Jesu Christi et sub pena duarum milium marcharum argenti debere inviolabiliter in perpetuum observari precipio et confirmo.'

by allowing them a free hand in the revision of the municipal laws. The legislation enacted during the Devotion flowed directly from the programme of reconciliation. It was natural that once the friars began to enforce their pacification by legal decree they would begin to provide for the future. The legislation in Parma, the remains of friar Gerard's reform there, is highly suggestive and thus demands a careful analysis.

8
The Revivalist as Legislator

THE combination of penance-preaching and reconciliation is an element of continuity in itinerant preaching from the early twelfth to the early thirteenth century. During the Great Devotion the friars added a novel element to this traditional programme, using the popular good will generated by their peace-making to gain appointment as reformers of municipal statutes and laws. In northern Italy before 1233, religious in general and the friars in particular had already played a direct role in ensuring the integrity of municipal governments. Treviso's statutes, compiled between 1231 and 1233, protected the security of city records by having them inscribed in two codices. One remained with the *podestà* of the city; the other lay sealed in the sacristy of the convent of the Friars Preachers. This storage box was locked, and the three keys were entrusted to the *podestà* and two other men.[1] The friars' involvement in maintaining the integrity of municipal governments should not surprise us. Their status as consecrated religious commended them. More than members of the older orders, such as the Benedictines and Cistercians, who were often powerful landlords, the friars seemed free from economic self-interest because of their vows of mendicant poverty. Furthermore, their mobility carried them to cities far from their place of birth. This transplantation disassociated them from family and clan and isolated them from local politics. To all but strong imperialists, who saw them as papal agents, they seemed safely neutral and above all faction.

The revivalists' direct involvement in politics extended beyond acting as a safe depository for documents. Some involvement was personal. It is reported—although when it occurred is unknown—that John of Vicenza's prestige and moral authority gave him the stature to confront armies in the field and order them to return home. All the revivalists were politically active in the legal process of peace-making. The preachers' most notable direct involvement in politics was their response to invitations from communes to take over control of the city governments and

[1] Trev. Stat. (344–8), 159.

reform their laws and institutions. The most lasting effect of this activity was their revision of municipal statutes, the results of which André Vauchez has recently studied.[2]

The friars' statutory reforms are, like the statutes themselves, a mostly uncharted world.[3] Although remaining municipal statutes for northern Italy, including those revised by the friars, have appeared, for the most part, in modern editions, they still await systematic analysis and study. Editors of municipal statutes, especially those of the nineteenth century, prefaced their editions with more or less adequate descriptions of the statute-books' contents and history. These studies are usually of restricted scope, each treating one city only. Historians of medieval law have generally ignored the statutes and focused their attention on the rediscovery and cultivation of the Roman Law. Here, modern legal historians have followed the prejudices of their medieval predecessors. To Odofredus, the thirteenth-century Bolognese civilian, the merchants and tradesmen who wrote and compiled municipal statutes were 'asses trying to make law'.[4] He thought the product of such legislation unworthy of serious study.

The dearth of comparative studies demands that any attempt to analyse the friars' legal reforms proceed with extreme caution. A short overview of the remaining thirteenth-century statute-books containing legislation by the friars gives us an idea of what evidence is available. Alleluia preachers may have reformed statutes or legislated in at least the following cities: John of Vicenza in Padua, Verona, Vicenza, and Bologna; Gerard of Modena in Padua and Parma; Peter of Verona in Milan; and Leo de' Valvassori in Monza. Henry of Cominciano, whom the narrative sources do not mention as an Alleluia preacher but Vauchez accepts as such, reformed the statutes of Vercelli, probably in 1234. John of Vicenza's reforms in Padua, Verona, and Vicenza, if they ever existed, have vanished. The same is true of the supposed reform by Peter of Verona in Milan.

---

[2] Vauchez, 519–49. On the relationship of peace-making to political power, see Harald Dickerhof, 'Friede als Herrschaftslegitimation in der italienischen Politik des 13. Jahrhunderts', *Archiv für Kulturgeschichte*, 59 (1977), 366–89. Dickerhof, 371–2, sees John's assumption of the titles of *dux* and *rector* as the natural outcome of peace-making.

[3] For a survey of statutory history, see Enrico Besta, in Pasquale del Giudice (ed.), *Storia di diritto italiano* (Florence, 1969), i (2). On the unsettled right of the cities to legislate for themselves, see Antonio Ivan Pini, *Città, comuni e corporazioni nel medioevo italiano* (Bologna, 1986), 141.

[4] Odofredus, i. 10. Boncampagno of Signi characterized non-professional legislators as 'idiote', cited in Besta, *Storia di diritto italiano*, 513.

The friars' surviving legislation is of very uneven quality and completeness. The evidence for John's activity at Bologna is very, very meagre. Although he made more or less extensive alterations in the Bologna statutes, only three of his acts can be identified with certainty in the extant 1245 codification. The three statutes that carry his name are all located in Book Five of the codification. Rubric Ten contains two statutes concerning Armanno of Porta Nuova, which have already been described. In Rubric Eleven, a peace pact between the same Armanno and the guardian of Gandulf's sons, one Juliana, completes the legislation on this dispute. There exists, besides these, only one other statute that is identifiably the work of friar John—the decree outlawing all sworn associations within the city of Bologna and its district. There are no other statutes securely traceable to John's legislative activity.

The hypothesis, which originated in the nineteenth century with Henry C. Lea, that the Devotion was above all a campaign against heresy led André Vauchez to speculate that some of the Bologna anti-heresy laws—in Rubric Eight of Book Five, the same book of the codex in which John's known legislation appears—belong to his reform. For Vauchez, John's peace-making provides a justification for identifying a series of acts against peace-breaking found in Book Two, Rubric Fifteen as his as well. To ascribe these laws to John is gratuitous. They are a type of legislation typical of most north Italian statutes from the period. There is simply no compelling reason to trace them to John rather than to some other Bolognese legislator of the period.

A block of laws enacted by the Franciscan, Henry of Cominciano, is preserved in the statute-book of Vercelli. Vauchez considered him a preacher of the Devotion.[5] Ten of his statutes deal with the suppression of heresy, three with the freedom of the Church, two with prostitutes, and one each with usury, the Studium Generale of Vercelli, and the communal debt. Laws written by Leo de' Valvassori for Monza exist, at least in part, in a codex edited by Anton Francesco Frisi in 1794.[6] These statutes deal entirely with heresy. Half of them merely repeat, virtually word for word, Henry's anti-heresy legislation at Vercelli. Leo altered only an occasional word or phrase to suit the particular circumstances at Monza. This legislation appears in a codex of documents dealing exclusively with the ecclesiastical affairs of the city. If Leo did legislate on non-ecclesiastical issues, these statutes have vanished. Since there is no

---

[5] Vauchez, 505. For texts, see Vercelli Stat., col. 1230–7.
[6] Monza Stat., 101–5.

evidence whether or not Leo enacted any other laws, these statutes could be very misleading as an example of the scope and thrust of Alleluia legislation in general.

Only for the reforms in Parma do we have sufficient material to make a secure analysis. The 1255 codification of that city's laws contains many traces of Gerard Boccabadati's reform of 1233. In this reform, he may have drawn on his earlier reforms at Padua,[7] but the condition of the Padua statutes excludes any certainty about this. The Padua statute-book of 1276 gives dates for statutes after 1236, but for those before that date it indicates only that the statute is an 'old statute promulgated before 1236' (*statutum vetus conditum ante 1236*). There is no way of isolating Gerard's work from that of any other pre-1236 legislator. This lack of specific attributions and dates also makes it impossible to locate any statutes by John of Vicenza in the Padua compilation. When all possibilities have been considered, a secure and comparatively complete example of the legislation exists for one, and only one, Alleluia preacher, Gerard at Parma.[8] Fortunately this evidence is ample. Any attempt to characterize the legislative activity of the friars must begin with this collection.

The oldest extant compilation of Parma laws contains forty-four laws or corrections of laws explicitly ascribed to Gerard. It is unknown how much of Gerard's legislation has disappeared. Some of his laws also contain later emendations. These emendations hint that at Parma, unlike Bologna, later revisers tended to preserve and emend earlier statutes rather than replace them with new ones. The Parma laws, unlike those of the other north Italian cities examined, also give the name of the legislator for a very large number of statutes. The Parma statute-book, then, is more a compilation than a systematic revision. For that reason it is likely that it preserves the bulk of Gerard's work. A brief statistical review of the Parma statutes is very revealing. Table 2 tabulates the forty-four statutes that carry the name of Gerard as legislator or reviser.

For Vauchez the friars' programme had six parts of roughly equal importance: suppression of heresy, defence of ecclesiastical liberties, moral reforms, peace-making, anti-imperial legislation, and restriction of the

---

[7] This is suggested by Besta, *Storia di diritto italiano*, 553.

[8] This reform was not by Guala of Bergamo, as Joseph Kuczynski, *Le Bienheureux Guala de Bergame de l'Ordre des Frères Prêcheurs* (Estavayer, 1916), 150, reports.

TABLE 2. *Gerard's Legislation at Parma*

Type of Legislation	Number of Statutes and Revisions
Peace and Justice Legislation	
Peace-making and peace pacts[a]	15
Control of the ban[b]	5
Protection of widows, wards, orphans, etc.[c]	5
Administration of justice[d]	2
The legal effects and status of the reform[e]	2
TOTAL	29
Ecclesiastical Legislation	
Against heretics[f]	6
Liberty of the Church[g]	3
TOTAL	9
Moral Reform Legislation	
Against fornication, adultery, etc.[h]	2
Against diviners, potion-makers, etc.[i]	1
Public decency[j]	1
Against blasphemy[k]	1
TOTAL	5

[a] Parma Stat., pp. 3, 216, 217, 221, 292 (3 statutes), 301, 302 (4 statutes), 305, 307, 312. [b] Ibid. 306, 307, 312, 313, 314. [c] Ibid. 5, 27, 289 (3 statutes). [d] Ibid. 3, 305. [e] Ibid. 199, 200. [f] Ibid. 10, 271 (5 statutes). [g] Ibid. 5, 198, 200. [h] Ibid. 43, 290. [i] Ibid. 42. [j] Ibid. 320. [k] Ibid. 319.

power of the magnates. Vauchez lays special emphasis on the importance of the suppression of heresy in the friars' programme.[9] A closer look at the topical breakdown of Gerard's legislation leads to a more nuanced description of the programme. Well over half the statutes, both in numbers and in bulk, deal with justice and peace. The ecclesiastical legislation does treat of heresy and the liberty of the Church, but even when taken together the statutes of these categories are only a third as numerous as those in the category of justice and peace. Finally, statutes on moral reform are only half as numerous as those treating ecclesiastical

[9] Vauchez, 549.

matters. Vauchez's last two categories, anti-imperial and anti-magnate legislation, do not appear at all, unless one could construe the three statutes to protect the liberties of the Church as anti-imperial. Surely these figures would imply varying levels of interest and activity by Gerard in his areas of legislation.

John of Vicenza's extant statutes (three about peace, one about civil order) also support this conclusion, as do the narrative sources. These universally emphasize the friars' roles as peace-makers and arbiters. Heresy suppression and moral reforms, while prominent, especially in the reports given by ecclesiastics, have at best a secondary role, if they have any at all. They are incidental or lacking in the lay sources. Finally, the narrative reports alone contain references to actions concerning usury, prostitution, and sumptuary laws.

Comparing the narrative sources with Gerard's laws suggests three prin-cipal areas of legal reform, labelled in Table 2 'peace and justice legislation', 'ecclesiastical legislation', and 'moral reform legislation'. These areas will be dealt with individually, beginning with the least important. The first, and in terms of quantity the least important, is moral reform legislation. Such reforms impressed observers but produced few statutes. The second area, of somewhat greater importance, is ecclesiastical legislation, that is, laws dealing with Church liberties, heresy, and the defence of religious and the weaker members of society. The last section, peace and justice legislation, which is of greatest im-portance quantitatively, includes that block of reforms aimed at removing causes of violence. This last section treats the friars' reforms of communal government itself and legislation intended to prevent abuse of power and ensure equity. These laws institutionalized the revivalists' peace-making and rendered it a part of the fabric of municipal systems of justice.

This analysis of the friars' legislative activity follows two principles: first, it allows the emphases of the narrative sources and the character and distribution of Gerard's extant legislation to suggest what was important to the friars in their work of reform; secondly, it tries to compare the effects of that legislation with legislation on the same matters found in the statute-books of other north Italian cities.

MORAL REFORM LEGISLATION

It has already been seen how a sermon by John of Vicenza could touch off a riot during which the house of Landolfo the usurer was burned, and the usurer himself only narrowly escaped lynching at the hands of an enraged mob. Anthony of Padua preached against usury during his Veneto campaign of 1232. Other Alleluia friars must have preached against usury. One might surmise that some part of the friar's legislation sought to control lending at interest. In fact, there is no anti-usury legislation definitely linked to an Alleluia preacher in the extant municipal statutes to confirm this supposition.[10] We find very little anti-usury legislation in any of the preserved thirteenth-century statutes for the region. Only two statutes of this type exist. They survive in the statute-books of Bologna and Vercelli. There is no concrete reason to link the Bologna statute to John of Vicenza; the Vercelli statute appears among the laws issued by Henry of Cominciano. These laws give us some idea of the kind of legislation the friars might have issued had they carried their anti-usury campaign beyond the stage of preaching and public agitation.

In the 1252 codification of the Bologna statutes, the usury law appears in a section of the statutes that prohibits the carrying of arms within the city, various ways of disturbing the peace, maltreatment of prisoners, gambling, and blasphemy. This block of legislation, considering its content, may well have come from an Alleluia preacher. There is, however, no evidence to that effect. The treatment of usury is interesting. It does not forbid all usury but only the granting of usurious loans to gamblers. In legislation on usury, then, this legislator did not act against usury *per se*. Rather, he attempted to defend the indigent—in this case those who had improvidently impoverished themselves through gambling—from being victimized by usurers. As a whole, these laws aim

---

[10] The anti-usury legislation discovered by Antonio Rigon in the medieval register of Alba, near Padua, may be the result of preaching by the Franciscan Henry of Padua in 1233, but there is no proof of a connection. On these laws, see Antonio Rigon, 'Francescanesimo e società a Padova nel duecento', in *Minoritismo e centri veneti nel duecento nell'ottavo centenario della nascita di Francesco d'Assisi (1182–1982)* (Civis studi e testi, 7; Trento, 1983), 17–21. I have found no evidence whatsoever that anti-usury agitation was directed against Jews. Sources for the Devotion do not mention them, whether as money-lenders or not. Prof. Gavin Langmuir of Stanford University, an expert on medieval anti-Semitism, has informed me that the absence of anti-Semitic preaching in this period is no surprise. He suggests that anti-Semitic preaching by friars became common only after the mendicants fell away from their original enthusiasm and suffered a consequent crisis of insecurity about their mission.

more at reform of public morals, here by the suppression of gambling, than at attacking usury in its own right.

Henry's legislation, if it does reflect the interests of the reformers of 1233, would also imply that anti-usury legislation played a secondary role in the friars' legislation. His statute legislates nothing directly against usurers or usury. Rather, in a single chapter, he suppresses two other statutes, which are consequently now no longer found among the laws of Vercelli.[11] He first voids Chapter 159 of the 1226 laws, which protected creditors from defaulting debtors. He then declares void the type of interest called *guidardonum*, which creditors exacted on loans of land. He also declares void a certain other law that was made to help usurers.[12] Here again, there is no attempt to legislate directly against usury. It seems, then, that the friars restricted their war against usury to public condemnation of usurers, organization of public resistance to their activities, and removal of their municipal protection or recognition. Perhaps in the legal realm the friars considered the canonical sanctions against usury sufficient.[13] Although it is dangerous to argue from silence, the absence of anti-usury legislation in the large corpus of laws that Gerard of Modena issued at Parma seems to confirm this.

The friars came to the aid of debtors. John of Vicenza ordered the release of those imprisoned for debt.[14] Gerard issued several laws to protect debtors.[15] When the slight extant evidence is carefully considered, it is best to describe the friars' anti-usury legislation as protection of debtors instead of a direct attack on usurers.[16] As such, their enactments fit better among the friars' attempts to remove the elements of division and hostility that underlay so many acts of urban violence, than in their programme of moral reform.

---

[11] Vercelli Stat. (ch. 383), col. 1236.

[12] 'capitulum quod faceret in auxilium usurarum similiter irritum sit'.

[13] For contemporary canonical treatment of usury, see X 5. 19, in particular, c. 3: bars manifest usurers from the sacraments, c. 5 and 9: X 2. 24. 6. compels usurers to return usurious gains. On this legislation, see Elisabeth Vodola, *Excommunication in the Middle Ages* (Berkeley, Calif., 1986), 129–31. As in Siena (William Bowsky, *A Medieval Commune: Siena under the Nine, 1287–1355* (Berkeley, Calif., 1981), 111–13), it is unlikely that the communal government vigorously enforced this legislation.

[14] Borselli, *Cron.*, 22.

[15] Parma Stat., 27.

[16] Carlo Sigonio, *Historia de Rebus Bononiensibus Libri VIII* (Frankfurt, 1604), 107, a rather late witness, confirms this when he says that John revised titles of credit to the benefit of debtors. In light of the statutes, Vauchez, 534, probably exaggerates when he says, 'La lutte contre l'usure apparait comme une des préoccupations essentielles des Mendiants.' The laws themselves give no evidence to support such a conclusion.

A second aspect of public morals that reportedly interested the preachers was the habits of women, both respectable and less than respectable. Revivalists' continuing interest in women predates 1233, as can be seen from stories about Foulques's reform of prostitutes and from the unseemly gossip about the popularity of young Dominicans with women. No legislation on prostitutes traceable to an Alleluia preacher survives.[17] Still, evidence that the friars' interest in women affected their legislation is suggestive, even if it is not plentiful. Borselli and the Bologna municipal chronicles report that John of Vicenza ordered women to wear veils. He also preached against frivolous displays in public dress, such as the wearing of crowns of roses. The Bologna statutes of 1253 reflect similar concerns when they carefully regulate the type of attire to be worn by prostitutes.[18] These statutes also restrict certain forms of adornment to honest women and others to prostitutes. Although these statutes carry no explicit mention of John of Vicenza, they are in accord with what is known about his preaching on dress. He urged the veiling of women. This practice, while probably suggested by the teaching of 1 Cor. 11: 10 that women cover their heads in church, served to set the devout apart from the disreputable, thereby clarifying the boundaries between the various elements in the society. That such identifying dress could prevent unintentional insults to respectable women is obvious. While John's primary intention was probably to follow the teaching of Saint Paul, his peace-making also benefited from the reduction of competition and discord brought by a standardization and simplification of dress. One might wonder how widely his hearers observed his directives.

Unlike that controlling usury and dress, some actual legislation against immorality and profanity has come down to us from the pen of an Alleluia preacher. Among Gerard's legislation there exist chapters against adulterers and 'diviners'—who also acted as pimps and dealt in poisons,

---

[17] Beyond regulation of the location of brothels, legislation against prostitution is found in the statutes of Henry of Cominciano, Vercelli Stat. (ch. 384), col. 1236–7: prostitutes living in a brothel ('in prostabulo') must move to a place outside the city and its district. Anyone who wishes may seize and rob those who do not. Finally the leaders of the commune shall flog any prostitute found in the city in the future. She is, however, to be let free after flogging. This statute forms part of a section against married men keeping mistresses in their houses. This legislation against prostitutes is thus a defence of the sanctity of marriage. *Assidua*, 344, tells that Anthony of Padua 'meretrices quoque a nephario prohibebat flagicio'. If he took any practical measures to reform them, these were not recorded.

[18] Bol. Stat., i. 309–13. This legislation also includes a crude drawing of the style of the dress and headgear to be worn by prostitutes.

contraceptives, and abortifacients.[19] Gerard's legislation against diviners
opens this way:

A chapter of Brother Gerard concerning a certain grave but very commonly
accepted crime in the city of Parma that harms and dishonours Jesus Christ and
the whole Christian people. Brother Gerard planned to act firmly and aggressively
in this case. Now, there were certain very evil wizards and enchanters, or, better,
lying betrayers, who are called diviners by poor ignorant men and women, and
who committed many crimes and terrible outrages against God that are shameful
before men. And these miserable wretches give poisonous draughts to stupid men
and women to commit evil acts, to seducers of women, and for destroying
children in the womb, and thus they lead men and women into evil acts and
serious crimes. They also receive fornicators and adulterers in caves and in their
houses under the pretext and hope of divining, and, after receiving payment from
them, they allow them and make them commit adultery, something that the
Most High Creator regards as horrible and ought to be so regarded by everyone.[20]

Gerard then decrees that the consuls and the *podestà* be bound by oath to
expel all diviners from the city within the first month of taking office.
Diviners subsequently captured were to be taken outside the city in
chains and severely flogged. A statute with a similar intent appears in the
Bologna statute-books. It immediately precedes a peace instrument issued
by John of Vicenza.[21] Although there is no direct evidence to connect this
legislation with John, it connects diviners (*divinatores et divinatrices*) and
sexual sin in the same way Gerard's statute did. They are placed under
the ban along with pimps, procurers, sodomites, public prostitutes, and
adulterers. Gerard placed adulterers under the ban in another Parma
statute.[22] He levied fines of £35 for a knight (*miles*) and £10 for a foot-
soldier (*pedes*) to remove a ban imposed for adultery. For the unmarried,

---

[19] On 'diviners', Parma Stat., 43–4; on adulterers, ibid. 290. On statutory repression of
magicians and sorcerers, see Edward Peters, *The Magician, the Witch and the Law*
(Philadelphia, 1978), 98–102.

[20] Parma Stat., 42–3: 'Capitulum fratris Gerardi super quodam gravi scelere et nimis
consueto in civitate Parmae ad detrimentum et ignominiam Jesu Christi ac tocius populi
Christiani. Cogitavit procedere frater Gerardus graviter et agresse. Sunt enim quidam
pessimi venefici et incantatores, immo potius falsissimi prodictores, qui divinatores a stultis
infoeneratis hominibus et mulieribus nominantur, facientes multa scelera et facinora graviora
Deo, et hominibus pudibunda. Dant enim isti miseri stultis hominibus et mulieribus pocula
venenosa pro iniquis factis, et deceptoribus mulierum et pro filiis in ventre destruendis; et
ut homines et mulieres ducant ad iniquos actus et scelera graviora. Et recipiunt fornicarios et
adulteros in speluncis et domibus suis sub praetextu et spe indivinandi, et pretio ab eis
sumpto, fornicationes et adulteria concedunt et faciunt exerceri, quod valde horribile videtur
Altissimo Creatori, et videri debet hominibus universis.'

[21] Bol. Stat., i. 446.   [22] Parma Stat., 290.

the cost of relaxation was only £10 for a knight and 100s. imp. for a foot-soldier, 'since those guilty of lesser crimes are to be punished with lesser penalties'.[23] In all cases the crime was to be made public ('facere cridare per civitatem infra mensem'), thus adding humiliation to the fine.

We know that John of Vicenza conducted a campaign not only against immorality but also against profanity. He taught the citizens of Bologna to respect the name of God and greet each other with the phrase 'Deus te salvet', 'May God save you'.[24] Anti-profanity legislation by the friars does not appear in the statute-books under consideration, probably because both Parma and Bologna already had statutes of this type on the books before 1233.[25]

Thus, although the documentary evidence is slight, it seems safe to say that, apart from adding new vigour to legislation against those sexual crimes that placed married life in danger, the moral reforms of the Alleluia preachers were more the result of their preaching than of their legislation. Moral reform was not unrelated to pacification. Like attacks on usury and regulation of dress, these reforms served to remove sources of division in society and threats to the stability of the family.[26]

ECCLESIASTICAL LEGISLATION

Henry C. Lea, Carl Sutter, and, more recently, André Vauchez have all seen the repression of heresy as a major, if not *the* major, component of the friars' statutory reforms.[27] Medieval observers appear to confirm the

---

[23] '[Q]uum minori poena puniendi sint minori crimine delinquentes.'

[24] Borselli, *Cron.*, 32; Rampona, 101. The greeting is clearly a blessing. On greeting-blessings as symbolic of exclusion or inclusion in a community, see A. van Gennep, *The Rites of Passage*, trans. M. B. Vizedom and G. L. Caffee (Chicago, 1966), 32. For laws on greeting in anti-heresy legislation, see Vodola, *Excommunication*, 7, 51–2.

[25] Parma's anti-blasphemy statute dates to 1228, Parma Stat., 332. It protected the name of God, the Virgin, and the saints and imposed a fine of 100 d. parm. or flogging. The same statute also prohibited games of chance on the steps of the Palazzo Comunale. Perhaps these were a cause of profanity. Bologna's blasphemy statute is likewise connected with gambling, see Bol. Stat., i. 299.

[26] There is one other piece of 'public decency legislation' in the statutes of Gerard. This is a law against defiling the walls of the cathedral, baptistery, or canons' residence ('turpe facere juxta murum majoris Ecclesiae. . .'), Parma Stat., 320. The fine was 3s. parm., or public display in chains in the Platea Communis.

[27] Henry Charles Lea, *History of the Inquisition in the Middle Ages* (3 vols.; London, 1886), ii. 203–5; Sutter, 10–12, 24–5, 36–8, 73–4; Vauchez, 523–7; this interpretation has also been recently adopted by Gary Dickson, 'The Flagellants of 1260 and the Crusades',

conclusions of these modern scholars. Stephen of Spain, during investigations for the canonization of St Dominic, reported:

In the cities of Lombardy a huge number of heretics has been burned, and more
than a hundred thousand people who did not know whether they ought to belong
to the Roman church or to the heretics have been sincerely converted to the
Catholic faith of the Roman church by the preaching of the Friars Preachers.
Their sincerity is shown by the fact that these converts, who had previously been
defending the heretics, are now hunting them down and detest them, and, in
almost all the cities of Lombardy and the Marches, the statutes that were opposed
to the Church have been handed over to the Friars Preachers to correct and emend
and bring into line with the Catholic truth.[28]

This description of the friars' activity does seem to imply that repression
of heresy and the institution of statutes against dissent were a major part
of the Alleluia programme. On the other hand, it was in Stephen's
interest to emphasize this element in his testimony, since his deposition
was made before papal judges. He was thereby linking the would-be saint
with the papacy's promotion of preaching against heresy and its desire to
have the anti-heresy decrees of Frederick II incorporated into the statutes
of the north Italian cities.[29]

I have great reservations about the tendency of scholars to characterize
the revival of 1233 as an anti-heresy crusade. The evidence for this
assertion is of three kinds. The first kind is circumstantial; revivalistic
religion in the early thirteenth century is usually tied up with attacks on
heresy. The second is narrative—the reports of John burning heretics
at Verona and Peter of Spain's testimony in particular. The third is
statutory—the existence of anti-heresy legislation by Brother Gerard at
Parma and Henry of Cominciano at Vercelli. While the first two types of
evidence are a compelling reason to believe that suppression of heresy did
occur during the Devotion, they tell us nothing about how important
this was in the context of the revivalists' programme as a whole. An
evaluation of the anti-heresy component of 1233 must depend on the
statutory evidence: here alone can we balance attacks on heresy against
other elements in the friars' programme.

Journal of Medieval History, 15 (1989), 247–8, who draws on N. J. Housley, 'Politics and
Heresy in Italy: Anti-Heretical Crusades, Orders and Confraternities', *Journal of Ecclesiastical
History*, 33 (1982), 193–8.

  [28] *Acta Can. Dom.*, 158–9; Quétif-Échard, i. 54, trans. from *Early Dominicans*, ed.
Simon Tugwell (Classics of Western Spirituality; New York, 1982), 81.

  [29] On these decrees and their effects see, G. de Vergottini, *Studi sulla legislazione imperiale
di Federico II in Italia: Le leggi di 1220* (Milan, 1952), 282.

The most powerful evidence for a characterization of the Devotion as an anti-heresy campaign is the reform legislation promulgated by Henry of Cominciano at Vercelli. These reforms would be a strong argument that the main thrust of the movement was anti-heretical, if Henry did enact them as part of the Alleluia.[30] There is no solid evidence that he did. That Henry was an Alleluia preacher remains to be proved. First, no contemporary source for the Devotion ever refers to him, by name or by implication, as a preacher of the movement. He would have received some mention in the writings of his fellow Franciscan Salimbene, at least. The only conceivable argument in favour of such a hypothesis is the date of his laws. The statutes carry no date, but Gregory IX, in a letter of 30 April 1235, upheld an interdict by the bishop of Vercelli against those who had violated the liberty of the Church protected by the statutes of a Franciscan friar named Henry.[31] Thus the statutes were enacted before that date. How much earlier, it is difficult to say. The statutes appear after other legislation, almost certainly not by Henry, which absolves Ardicino Blandrato (Stat. 388) and Calderia and Buongiovanni Ferri (Stat. 389) from bans incurred during factional strife during the years 1234–5. The presence of Henry's work after those statutes led the editor of the Vercelli statutes, Giovambatista Adriani, to date the Franciscan's reform to 1234, the year after the Alleluia.[32] Henry's identification as an Alleluia preacher rests on that dating and on two known facts, that he reformed statutes and was, like Gerard, a Franciscan.

Certainty is impossible, but the character of his laws would imply that Henry acted, not as a general corrector of statutes, as John did in Bologna, but simply as an ecclesiastical agent to introduce anti-heresy statutes into the Vercelli laws. The preamble to his work tells us that Henry's legislation was 'to the honour of the Holy Roman Church, our Mother, and of the Lord Pope Gregory, and the venerable Lord Hugo, by the grace of God, bishop of the holy Church of Vercelli, for the defence of the Catholic Faith'.[33] The Emperor Frederick's decree of March 1224, the 'Constitutio in Basilica Sancti Petri', appears among these statutes. This imperial legislation provided for a forceful suppression of heresy, so Henry merely needed to adapt it to local conditions and needs. The grouping of all Henry's decrees in one place, seemingly as a single act of

---

[30] Vauchez, 523–6.   [31] Vercelli Stat., col. 1229, n. 95.

[32] Besta, *Storia di diritto italiano*, 513, gives the date as 1234.

[33] Vercelli Stat. (ch. 369), col. 1231, '[A]d honorem Sacrosancte romane Ecclesie et matris nostre, et domini G. summi Pontificis, et domini Hugonis Dei gratia Sancte Vercellarum Ecclesie venerabilis Episcopi, et ad defenssionem catholice fidei.'

legislation, also goes against the practice of the Alleluia preachers who 'added, corrected and suppressed the laws according to their own will', to use the phrase describing the activity of John of Vicenza. In short, the form of Henry's activity differs markedly from that of the activity of known Alleluia preachers like Gerard and John. To understand the role of heresy legislation in the friars' reforms, it is safer to base the analysis on the evidence provided by a preacher like Gerard, who was certainly of the Alleluia.

Heresy legislation must be seen in context. To the extent that it was an action against heresy, the Alleluia was part of a process that extended from the 1220s to the late thirteenth century.[34] Cardinal Ugolino began it in 1221, when he had the 1220 imperial law making heresy a civil offence entered into the statutes of Mantua, Bergamo, and Piacenza.[35] In 1224 the Emperor Frederick declared death by burning to be the imperial punishment for heresy. Treviso appears to have adopted this legislation against heresy in 1228, creating committees to enforce it.[36] Verona adopted similar legislation in 1228, revising it later in 1272.[37] Vauchez, citing the testimony of Corio, would have it that Milan accepted the imperial legislation on heresy in 1229.[38] Later, the imperial acts against heresy were accepted, at Monza in 1233 or 1234,[39] at Vercelli in 1234, at Bergamo in 1239, at Padua in 1239, and at Bologna in 1239.[40] Apparently, formal acceptance of the imperial legislation did not occur at Parma until 1261, although other municipal acts against heresy already existed before Gerard of Modena revised the city's statutes in 1233.[41]

Two facts stand out in this review of anti-heresy legislation. First, in the fragmentary statutory remains of the north Italian cities, the enactment of the imperial statutes and other anti-heresy laws appears as a

---

[34] On willingness to persecute heretics in north Italian cities, see Antonio Rigon, 'Chiesa e vita religiosa a Padova nel duecento', in *S. Antonio 1231–1981: Il suo tempo, il suo colto, e la sua città* (Padua, 1981), 293.

[35] Christine Thouzellier, 'La Légation en Lombardie du cardinal Hugolin', *Revue d'histoire ecclésiastique*, 43 (1950), 524–6.

[36] Trev. Stat. (634), 250.

[37] Ver. 1228 Stat., 116, with revisions, ibid. 202.

[38] Vauchez, 524, quoting Bernardino Corio, *Storia de Milano* (3 vols.; Milan, 1855–7), i, 408–9.

[39] This Monza legislation by Leo de' Valvassori is partly identical to Henry's. Perhaps both reflect a systematic Franciscan effort against heresy in Lombardy.

[40] The Monza laws are cited by Vauchez, 535. For the other cities see: Vercelli Stat., 234–5; *Antiquae Collationes Statuti Veteris Civitatis Pergami*, ed. Giovanni Finarzi (Historiae Patriae Monumenta, 16: Leges Municipales, II; Turin, 1876), cols. 1934–6; Padua Stat., 423; Bol. Stat., i. 464.

[41] Parma Stat., 272.

protracted process, extending throughout the 1220s and 1230s; so even if a friar enacted anti-heresy legislation, this did not mark him as distinct from other communal legislators. Second, it is striking that, except for Monza, there is no evidence for the introduction of anti-heresy laws, or the introduction of the punishment by fire, in any north Italian city during the year 1233, the year of the Great Devotion. If we again exclude Leo's reforms at Monza, there is no extant example of the introduction of imperial anti-heresy legislation by a known Alleluia preacher. Although the city laws are not preserved in their entirety, what remains suggests that the friars' activity was limited to reform of the existing anti-heresy laws, not the enactment of new ones.

Gerard's legislation at Parma confirms this conclusion. Parma already had a comprehensive anti-heresy statute at the time of Gerard's reforms in 1233.[42] This law bound the *podestà* and the rectors of the city to punish within eight days anyone declared a heretic by the bishop of the city; likewise, they were to punish any Cathar they might discover. The punishments were to be 'severe', so as to serve as an example to others. The law also set up committees and procedures to assist the bishop in this activity. The *podestà*, under the direction of the bishop, was to establish a board of four men to root out secret preaching or meetings and to uncover heretics, their adherents, and their supporters. Other government officials were to grant the four full liberty of action, under penalty of a fine of £10. In his investigations, the *podestà* was bound to apply torture to suspects at the will of the bishop. This decree came well before the approval of the use of torture in ecclesiastical courts in 1252.[43] Those found guilty of heresy were to be publicly named and fined £10 parm.; persistence in heresy, or its continued defence, doubled the fine. The guilty were gaoled at the will of the bishop if they did not pay. Fines were also levied on those who housed or helped heretics (£20 parm.) and on those who attended their sermons (£3 parm.). No future reviser of the communal laws had the power to change or diminish the legislation, and the statute bound the *podestà* to exact an oath from his successor to enforce the penalties. In cases of doubt, the bishop had the power to interpret the law.

These acts carry no date, but the early 1220s is a likely guess since they reflect the imperial legislation of 1220.[44] Notably absent is the

---

[42] See Parma Stat., 269–71.
[43] By Innocent IV in *Ad Extirpanda* (Potthast 14,592).
[44] Vauchez, 527.

punishment by fire decreed in 1224. Even more interesting is Gerard's failure to include capital punishment in his five additions to the heresy statute.[45] Three of these do little more than repeat the extant legislation: The *podestà* and consuls are again required to enforce the law. The bishop's right to rule in dubious cases and his right to order torture are reaffirmed. Finally, the statutes are again declared perpetual and immutable. Only two new measures are enacted: first, Gerard declares public and private disputations against the Catholic faith the equivalent of heresy and imposes a fine of 100s. imp. on knights and one of 50s. on foot-soldiers for violation of this prohibition; secondly, the statutes are to be read publicly three times a year. There is only one other act by Gerard about heresy in the compilation; his declaration of heresy as a bar to holding public office.[46]

Before evaluating the significance of Gerard's anti-heresy legislation, it would be useful to contrast Parma's heresy laws with those of other north Italian cities of the same period. Municipal laws roughly analogous to those of Parma exist for Treviso, Vercelli, and Verona before 1233.

The Treviso legislation is roughly similar to that in force at Parma in 1233.[47] It requires that citizens report any heretics they discover (§1), that the heretics' houses be destroyed (§2), that their supporters be fined (§3–4), and that a committee of two Catholics from each quarter be established to suppress heresy (§5). It also outlines procedures to be followed (§6). Unlike the Parma statutes, however, Treviso's provided that the relapsed be punished 'according to the imperial laws', that is, with burning (§7), and that heretics' children and grandchildren be deprived of any right to ecclesiastical benefices (§7). As a whole, this legislation, which John of Vicenza would have found in force in 1233, is more severe than that of Parma. The major difference is that at Treviso there is no explicit role for the bishop in the fight against heresy. The Treviso statutes treat heresy as a civil crime.

Vercelli's 1234 revisions by Henry of Cominciano are the most elaborate anti-heresy laws found among extant north Italian statutes roughly contemporary with the Alleluia. They order: the expulsion of heretics, the organization of groups of men under the direction of the *podestà* and bishop to seize heretics, the exclusion of heretics from public office, prohibitions against aiding heretics or providing them with shelter, and the public placing of discovered heretics under the ban

---

[45] Parma Stat., 271.   [46] Parma Stat., 10.

[47] Trev. Stat. (634), 250. The editor of these statutes, Giuseppe Liberali, suggests a date of 1228 for this law, a not unreasonable suggestion.

(§370). Specific statutes punish those who harbour heretics (§371) or prevent their capture (§372). Several statutes provide for oaths: by officials that they are not heretics (§383), by electors that they will not knowingly elect heretics (§384), and by the *podestà* that he is not a heretic (§375) and that he will enforce anti-heresy legislation (§378). Finally, rules of procedure are envisioned (§377), although here there is no mention of torture. In conclusion, the text of Frederick's decree of 1224 appears appended to the section. This legislation does not differ notably from what is found elsewhere. Henry's work is simply more systematic and elaborate.

The anti-heresy statute in force in Verona was brief and has the form of an oath by the *podestà*. It reads in part as follows:

> I shall expel heretics and Patarins from the city and its district, unless they have come at the will of the Lord Bishop, nor shall I permit them to stay there: all this at the command of the Lord Bishop. And the house or houses in which they live, I shall destroy or have destroyed . . .[48]

Here, the law specifies no punishment other than exile, but it provides for co-operation between the bishop and *podestà* similar to that in Parma. Since this was the only anti-heresy law in force in 1272, unless some other has disappeared, there is no evidence that John made any revision during his residence in the summer of 1233. If, as Parisio says, he took the opportunity, contrary to the extant Verona statutes (which do not envision such a penalty), to burn sixty Veronese heretics of both sexes at that time, this might suggest that he introduced capital punishment in a statute that has perished.[49] This execution is the only known occasion where a preacher is said to have burned heretics during the Alleluia. As *rector* of the city and equipped with a pontifical mandate, John did not need a municipal law; he might have acted on his own in this affair.[50]

---

[48] Ver. 1228 Stat. (ch. 156), 117–18: 'Et Haereticos et Patarenos expellam de Civitate, et eius districtu, nisi venerit ad voluntatem Domini Episcopi [vel eius vicarii], nec morari permittam: haec omnia ad praeceptum Domini Episcopi [vel eius vicarii]. Et domum sive domos, in qua, vel in quibus morabuntur, destruam, vel destrui faciam, si ille, cujus fuerit domus, ipsos tenuerit post octo dies a denunciatione sibi facta, vel postquam fuerit denunciatum in Concione per me vel meum Nuncium, ne nullus teneat Haereticos, at postea 15 diebus elapsis repertus quis fuerit eos in domo, vel domibus suis teneri, vel morari passus fuerit, eas domos destruam, vel destrui faciam a me, vel a meo Nuncio, sive Officiali.' The bracketed phrases were added during the revision of 1272.

[49] Parisio, 8: '[D]ictus frater Ioannes in tribus diebus fecit comburi et cremari in foro et glara de Verona 60 ex melioribus inter masculos et foeminas de Verona, quos ipsos condemnavit de haeretica pravitate.'

[50] As mentioned earlier, I suspect that the necessities of peace-making motivated this burning. Peace-making required routine use of oaths. The Patarins would have refused them

The content of Parma's anti-heresy laws, left virtually unchanged by Gerard, is almost perfectly mirrored in the laws of the other three cities. Verona envisions close co-operation between the bishop and the commune and takes action against those who house heretics. Treviso sets up a public committee to enforce the laws. The major difference is that in Parma the laws are more extensive (at least in comparison with Verona) and more systematic. On the other hand, they are far less severe than those of Treviso, which would, by implication at least, inflict the death penalty. There is little of substance to distinguish Gerard's laws from those of Henry. In conclusion then, Parma's anti-heresy statutes, in the form in which Gerard left them, are not markedly different from similar legislation that had already appeared elsewhere in north Italy. It seems safe to assume that the friars restricted their legislation on heresy to minor emendations and attempted to foster more rigorous application of the existing laws. So this examination of the laws themselves confirms what the quantitative analysis implied: as legal reformers the friars paid only secondary attention to heresy.

A more careful rereading of Stephen of Spain's testimony would also support this conclusion. He tells us that the friars' activity brought about conversions from heresy not by legislation but by preaching. Furthermore, the harrying of heretics (and, most likely, the burnings, where they occurred) seems to be the result more of the converts' enthusiasm than of direct legislation by the preachers. Finally, he says that the friars revised city law-books to bring them more into line with the Catholic truth. This could imply many types of new statutes besides those against heresy. On that score, the friars did not have to draft anti-heresy laws, because usually such laws already existed. So, although the Devotion of 1233 did include the reassertion of orthodoxy against heresy, it cannot be characterized as an anti-heresy crusade.

André Vauchez, in agreement with Lea and Sutter, considers the freedom of the Church as a major goal of the Alleluia preachers.[51] He brings as evidence for this a single statute issued by Gerard of Modena and entitled 'De Statutis contra Libertatem Ecclesiae'.[52] This statute commands each

for religious reasons. Parisio, 8, says that the heretics were 'ex melioribus inter masculos et foeminas de Verona', and thus from the very families John of Vicenza wanted to reconcile. That the burning of heretics may have also served to facilitate peace by removing a source of division within the community is suggested by Daniel A. Brown, 'The Alleluia: A Thirteenth Century Peace Movement', *Archivum Franciscanum Historicum*, 81 (1988), 12.

[51] Vauchez, 528–32.

[52] Found in Parma Stat., 198: 'Capitulum Fratris Gerardi quod Potestas Parmae teneatur per se vel per Judices suos infra III. menses, postquam habuerit Statuta in suis viribus et

podestà to examine the law-books of the city and within 3 months of entering office to cancel any laws contrary to the liberty of the Church. Perhaps of equal importance in this statute is the declaration that Gerard himself did not undertake a careful investigation of this type, leaving that task to future *podestà*. Nevertheless, it is not impossible that Gerard cancelled some statutes contrary to the liberty of the Church during his revision. Still, the absence of any other legislation by him dealing with ecclesiastical rights confirms his statement that he had left the systematic defence of Church liberty to future *podestà*.

Gerard did issue two other enactments on Church liberty, and they are worth a glance. One of these is a brief statute commanding the *podestà* and the consuls to defend the rights of the 'Brothers of Penance' in the city.[53] The second is a modification of the *podestà*'s oath.[54] His other additions to the oath extend well beyond the defence of the Church establishment. He binds the *podestà*, his judges, and his knights, first, to maintain the rights of the bishop, his clergy, churches, and hospitals, and the rights of all orphans, widows, and any individual, small or great, without respect of persons.[55] A pledge to maintain the rights and jurisdictions of all religious institutes or ecclesiastical corporate persons follows. The additions, then, extend beyond the rights of the clergy and religious institutes to those who were traditionally considered in some way under the protection of the Church because of their helplessness.[56]

capitula Statutorum, teneatur expressim ex debito sacramenti inquirere diligenter et perscrutari totum volumen Statutorum, et singula capitula invenire, et si invenerit aliquod Statutum in volumine, seu voluminibus nominatis, quod sit contra libertatem ecclesiae et ipsius debitam rationem, illud capitulum et Statutum [debeat cancellare?]. . . . Fratris Gerardi in sua propria persona inquirendis et perscrutandis ipsis Statutis non potuit tam subtili indagatione adesse, ex tunc ea omnia generaliter, si qua sunt vel inventa fuerint, ut superius continetur, cassavit et anihilavit, et nullius momenti pronunciavit, sicut superius statuit cancellanda: et de hoc non possit peti parabola, neque dari, quin fiat infra terminum supradictum.' The text appears to lack a line; 'debeat cancellare' is the editor's suggestion.

[53] Parma Stat., 200. Vauchez seems to have overlooked this text. The 1249 revisers of the Parma Statutes added a specification to this statute to clarify that it gave 'fratres commorantes in domibus eorum cum familiis eorum' no exemption from military service. Later revisers did not feel bound by Gerard's legislation to defend every sort of ecclesiastical exemption.

[54] Parma Stat., 5. On this text, see Vauchez, 529.

[55] 'Potestas et judices et milites Potestatis facient rationem, et manutenebunt jura primo Domino Episcopo, et clericis suis et ecclesiis et hospitalibus, orphanis et viduis et omnibus aliis personis tam parvis quam magnis sine acceptatione aliquarum personarum.'

[56] Gerard's legislation is similar to that at Vercelli by Henry of Cominciano, Vercelli Stat. (§380–2), cols. 1235–6. These statutes protect the liberties of the Church (§380), forbid statutes contrary to it (§381), and, unlike Gerard, protect privileges granted by the Roman pontiff (§382). More lengthy and comprehensive than Gerard's, these are only a

The protection of weaker members of society—a grouping that, for the friars, included members of religious orders—deserves some attention in its own right. In Gerard's reform the number of statutes protecting the weak is equal to that of those protecting the liberty of the Church—three in each case. The statutes protecting the weak are considerably longer and more elaborate. This suggests that such legislation played a significant, even if secondary, role in the friars' programme.[57]

Gerard was not satisfied to include the obligation to protect widows and orphans in the *podestà*'s oath alone.[58] He also bound each *podestà* to select, within one month of his taking office, a board of 'four good judges' to serve without charge as advocates, counsellors, and patrons of widows, orphans, and other 'miserable persons'.[59] He explicitly defined such persons as any who lacked the personal resources to support themselves.[60] The provision binds these judges under oath to serve all comers kindly and without deceit. An addition of 1242 granting a salary of £3 parm. to these 'patrons of the poor' (*patroncinatores pauperum*) proves that this law did not remain a dead letter. There is a similar concern for providing 'legal aid' in a statute from Padua.[61] Here guardians (*tutores pupillorum*) specifically receive power to bring civil and criminal action against those who have murdered the parents of their wards. Perhaps this statute originated during Gerard's Padua reforms; it does reflect the concerns found in his Parma legislation.

Gerard's interest in widows and orphans extended beyond their legal protection, into what is today called family law. For example, Gerard tightened up the protection of family rights in the marriage of daughters, by imposing penalties on those who married an orphan without the consent of her mother or brothers, thus creating a law analogous to one that already punished the marrying of daughters without the consent of

small part of Henry's legislation, nearly all of which is directed against heresy. I would conclude then that, even in the mind of a reformer with as narrowly ecclesiastical an interest as Henry, liberties of the Church came well after heresy in importance.

[57] Daniela Gatti, 'Religiosità populare e movimento di pace nel'Emilia del secolo XIII', in *Itinerari storici: Il medioevo in Emilia* (Carpi, 1980), 99, who does not know of Vauchez's work, lays heavy emphasis on this aspect of Gerard's programme.

[58] Parma Stat., 5.

[59] Parma Stat., 27: 'Potestas teneatur . . . quatuor bonos Iudices et legales eligere et sapientes de melioribus huius terrae, quibus salarium certum constituatur a Conscilio Parmae, qui debeant esse advocati et consciliarii et patroni orphanorum et viduarum et miserabilium personarum sine aliquo precio vel munere ab eis recepto.'

[60] '[H]oc intelligatur in illis orphanis et viduis et pauperibus qui vel quae per suam inopiam et indigentiam et paupertatem solvere non possunt.'

[61] Padua Stat. (715), 240. This law is dated 'ante 1236'.

their fathers.[62] In an addendum to the same statute, he extended this
provision to the marriage of males under the age of 16, and, in defect of
parents, he provided for the selection of relatives from the paternal and
maternal families to act *in loco parentis*.

Gerard seems to have directed this legislation principally towards the
preservation of family harmony by specifying legal norms for the sensitive
area of marriage arrangement and negotiation.[63] A description of
Ambrose Sansedoni's legal agitation found in his vita confirms this as an
interest of the revivalists:

[Ambrose] showed the greatest charity toward children and widows in the cities,
and he sought laws in those things that concerned them, providing for their
defence, and suitable persons were chosen for the defence of their goods and
prevention of injury, and, if word of injury to children or widows came to him,
he prayed to the Lord for their rights.[64]

What is the general picture presented by Gerard's moral and
ecclesiastical legislation? A careful reading of the extant texts, along with
their rarity, implies that these areas of legislation played a secondary role
in the friar's reforms. Furthermore, one would best understand them
as complementing and supporting the major thrusts of his preaching-
campaign, that is, the prevention of violence and the resolution of
conflict. The morals legislation, such as that against usury or diviners,
seems intended to do more than extirpate vice. It also served to preserve
the unity of the community by, for example, removing opportunities for
adultery and clarifying the positions of 'honest' and 'dishonest' women
within the fabric of the society. Similarly, Gerard's legislation on heresy
seems to aim particularly at co-ordinating the community's efforts to
suppress it, to make the programme more uniform and less haphazard.
He was quite willing to leave the suppression itself up to the community.
This was, no doubt, a far less time-consuming solution than prosecuting
heretics himself. Gerard's legal protection of the weak and his revision of
family law exhibit the same concern for promoting the smooth function-
ing of society. In short, Gerard's moral and ecclesiastical legislation

[62] Parma Stat., 289.

[63] As Rolandinus, fo. 140^r, reminds us, marriage arrangements served as a means to
preserve peace agreements. Thus, Gerard's interest in marriage law is not surprising.

[64] Alessandrino, 183: 'Maxima etiam erga pupillos et viduas usus est caritate in
civitatibus, in quibus eum esse contigisset, procurando quod pro eorum defensione leges
conderentur, eligerenturque personae ad hoc idoneae, quae circa bonorum ipsorum
defensionem ac injuriarum illationem eis essent praesidio, et si forte pupillis et viduis
molestiae illatae ad ejus pervenissent notitiam, pro eorum justitia Dominum precabatur.'

complemented and enhanced the promotion of peace and the elimination
of the urban conflicts and divisions that the preponderance of the statutes
and the testimony of contemporaries tell us were the Alleluia preachers'
primary concern.

PEACE AND JUSTICE LEGISLATION

The legal and social realities of the friars' role as public mediators and
arbiters have already been made clear. It is now possible to summarize the
legislation by which the friars sought to perpetuate the fruits of their
peace-making and to prevent future public unrest and urban violence.
Along with measures to keep those reconciled faithful to their peace
agreements, the friars also took steps to head off future violence.

The primary legal responsibility for preventing violence and ensuring
justice in the north Italian cities usually lay with the *podestà*. In Parma, as
his oath shows, this was the *podestà*'s primary responsibility.[65] In an
addition to his oath (which was probably drafted by Gerard), the statutes
themselves bound the *podestà* to sit in session with the bishop during each
week of Lent and assist with arbitration and reconciliation.[66] Gerard
enacted that all reconciliations brought about by him were to be enforced
and maintained inviolate by the *podestà* and officials of the commune as
part of their activity as public peace-keepers.[67] It appears that communal
legislators normally confirmed the pacts concluded during the *podestà*'s
sessions, as well as those issued by other arbiters, by enrolling them
in the city laws. Gerard had his acts entered into the statute-books
of Parma.[68] Even if the texts of peace accords eventually disappeared
from the statute-books, they continued in force and were occasionally
reaffirmed in public. Gerard's were solemnly reaffirmed twenty-seven years
later in 1261, and the reconciliations of John of Vicenza were reaffirmed
thirty-one years later in the 1262–4 revision of the Bologna statutes.[69]

Many circumstances limited the *podestà*'s ability to maintain the peace.

---

[65] Parma Stat., 3–4. Vic. Stat., 14–15, give a similar responsibility to supervise
reconciliations and mediate disputes to the *podestà* of Vicenza.

[66] Parma Stat., 3: 'Et quod teneatur Potestas suo sacramento in qualibet hebdomada
Quadragesimae majoris usque ad Pasca resurrectionis esse cum Domino Episcopo Parmae et
super esse et tractare cum eodem de pacibus et concordiis faciendis, et eas bona fide et sine
fraude facere et ad finem producere.'

[67] Parma Stat., 301.   [68] Parma Stat., 199–200.

[69] For Gerard, Parma Stat., 221; for John, Bol. Stat., i. 267. There seems to have been
considerable anxiety about pacts falling into obscurity or being ignored. The reaffirmation at
Bologna of 'paces et concordie facte tempore Devotionis' specified that the *podestà* reread

In theory, at Vicenza for example, the citizens were under oath to help him maintain the peace and put down revolts.[70] The same was true in Parma, where the *podestà* could call on the citizens to help put down riots.[71] Nevertheless, it is hard to believe that citizens saw their oaths as requiring them to take up arms against their own relatives. In Bologna, for example, the Società delle Armi, which were above all peace-keeping forces, specifically exempted their members from having to take up arms against 'fathers, brothers, or blood relatives.'[72] The *podestà* could enforce the peace only if he received the support of a large proportion of the population and a means existed for them to co-operate with him. It seems likely that a major reorganization of the Società delle Armi was carried out in 1233. It may well be the case that John assisted in this process.[73] These societies were especially attached to him.[74] If this conjecture is correct, John had found another, non-statutory, way to help the commune perfect a rudimentary police-force that could enforce the peace.

The typical *podestà* had, at least on paper, the power to impose punishments on those who broke truces or the peace. In Parma, for example, statutes bound all under penalty of law to inform the *podestà* of anyone who had sworn to help break a peace.[75] A 1228 law of Parma gave the *podestà* the power to impose capital punishment and mutilation on peace-breakers.[76] In Verona, the statutes simply required the offender, under penalty of the ban, to do whatever was necessary to be reconciled to the

them publicly and that this be repeated every three months! That cities reaffirmed the friars' reconciliations over 30 years after mediation does not harmonize with the opinion that the friars' peace-making had no real, enduring effects.

[70] Vic. Stat., 70, in the 'Sacramentum communitatis populi vicentini'.

[71] Parma Stat., 217–18.

[72] Bol. Armi Stat., 122, for the oath of the Balzani (1233) exempting 'patribus, fratribus, et consanguines'.

[73] So Augusto Gaudenzi, 'Gli statuti delle Società delle Armi del Popolo di Bologna', *Bollettino dell'Istituto Storico Italiano*, 8 (1889), 20, whose argument is conjectural: 'E siccome poi abbiamo memoria di uno statuto dei Toschi del 1233 e possediamo una matricola della società della Branca di porta Castello di questo stesso anno, e di più sappiamo dal citato statuto del Popolo del 1248 che nell 1233 fu rinnovato il giuramento degli uomini delle società delle arti e delle armi, e del cambio e della mercadanzia ciò che accenna di per sé a una riorganizzazione dei questi statuti accaduta dopo il nuovo moto populare del 1232, io soppongo che nell'anno 1233 sia avvenuto quest rimaneggiamento d'una divisione fatta prima.'

[74] Borselli, *Cron.*, 22; Rampona, 102.

[75] Parma Stat., 291.

[76] Parma Stat., 291: 'Rector civitatis Parmae teneatur quod si aliquis ruperit vel rumpere fecerit pacem alicui de districtu Parmae interficiendo vel magagnando, quod Potestas teneatur eum bona fide capere, et omnem operam dare ad capiendum ipsum etiam expendendo de avere Communis; quo capto, teneatur ei amputare caput in Concione coadhunata campana pulsata. Si vero no interfecerit . . . teneatur Potestas . . . ipsum capere et manum amputare in ipsa Concione.'

family of the injured party.[77] An indirect way to prevent feud and peace-breaking was to destroy an individual's ability to create the web of armed supporters that made it possible. Family and clan could not be legislated out of existence, but private sworn societies could. John specifically outlawed all such groups and declared their oaths void.[78] Such a law not only rendered illegal those groups pledged to defend each other or fight together; it also struck at the practice of receiving oaths of loyalty from retainers. Although no other legislation of this sort exists from the pen of an Alleluia preacher, such laws had entered or would enter the books of Parma, Verona, and Vicenza.[79]

Capital punishment or mutilation was not necessarily the best solution to the problem of urban violence—the shedding of blood tended to provoke further bloodshed. The best solution was the reconciliation of the parties. This seems to have been the intention of Gerard's revision of the punishment for peace-breaking. Contrary to what might be expected, Gerard's handling of the punishment for peace-breaking had the effect of relaxing earlier harsh penalties. He revoked the older capital punishment by declaring exile and the ban to be the punishment for those who broke the peace or caused it to be broken.[80] This seems part of a general movement to mitigate unnecessarily harsh punishments. A 1233 Parma statute found with those of Gerard, for example, protects those banned for debt from further harm.[81] Such protection seems to be part of an attempt to make the punishment of peace-breakers more equitable and even-handed. To this end Gerard ordered that definitive lists of those under the ban for peace-breaking be compiled and read periodically in public to prevent names from being surreptitiously added or dropped.[82] He also bound the *podestà* under oath to maintain and enforce all these bans.[83]

A reform of the procedure for imposition of the ban went only part way towards the restoration of peace and unity within the commune. It was also necessary to systematize the relaxation of the ban and provide for the

---

[77] Ver. 1228 Stat. (ch. 86), 67–8.   [78] Bol. Stat., i. 262–4.

[79] Against sworn 'conjurae' and 'conjurationes': Parma Stat., 176. and Vic. Stat., 130; against accepting vassals: Ver. Stat., 133; Parma Stat., 220 and (for 1252) ibid. 338; and Vic. Stat., 130.

[80] Parma Stat., 292. Henry of Cominciano mandates the same punishment for peace-breakers, Vercelli Stat., col. 1126.

[81] Parma Stat., 314: '[B]annum non debeat ei nocere quominus possit.'

[82] Parma Stat., 307–8. For a roster of those under the ban for peace-breaking and furtive murder, probably compiled in accord with this law and bearing the date 1239, see Parma Stat., 313.

[83] Parma Stat., 308–9.

return of exiles. Two statutes by Gerard, issued as a preparation for his peace-making at Parma, give us some idea of how this happened. Treating the relaxation of the ban:

[Gerard] decreed that the Podestà is bound to exact suitable security of keeping, preserving, and fulfilling the accords made and to be made and the restrictions and conditions given or to be given on the occasion of peace-making, from those by whom they have been requested and from those from whom they have not been offered. And if anyone does not make peace or give security at the will of the Podestà, the Podestà is bound to place him under the ban and not to release him from the ban unless he first make peace and give security according to the will of the Podestà. And that one who seeks peace, either for himself or another, is to be released from the ban in the manner prescribed above, that is, by giving security.[84]

Gerard also provided special treatment for those seeking absolution during the period of his peace-making, from the third Friday of July until the Feast of Saint Michael. He provided for exiles in a statute that reads in part:

Brother Gerard also determined that if anyone was not able to enter the city of Parma because of exile imposed on him, any person might be heard on his behalf who wished to have him freed from the ban, if the ban was of the type from which he could be released.[85]

If Gerard's earlier success in Modena is indicative, many took advantage of the times of amnesty.[86] John of Vicenza routinely ordered the release of prisoners; declarations of amnesty were a regular part of the Alleluia peace campaigns. This tradition of clemency as a road to communal peace was to endure. When the Dominican Remigio de' Girolami wrote a small tract on communal peace in the late thirteenth century, he saw the remission of punishments and forgiveness of injuries as essential to any peace-making process.[87]

---

[84] Parma Stat., 304: 'Item [Gerardus] statuit quod Potestas teneatur dare facere ydoneas securitates de pacibus factis et faciendis, et de confinibus de conditionibus datis et dandis occasione pacium tenendis et conservandis et adimplendis illis a quibus fuerint petitae ab eis a quibus praestitae non sunt. Et si aliquis pacem non fecerit et securitatem ad voluntatem Potestatis non praestiterit, Potestas teneatur eum in banno ponere, de quo banno extrahi non possit nisi primo pacem fecerit, et securitatem ad voluntatem Potestatis praestiterit. Et ille, qui pacem pecierit per se vel per alium, de banno extrahatur secundum formam superius ordinatam, scilicet de securitate praestanda.'

[85] Parma Stat., 313: 'Item statuit Frater Gerardus quod, si aliquis non possit venire in civitatem Parmae occasione confinium ei datorum, quaelibet persona audiatur pro eo volens eum facere extrahi de banno, si tale sit bannum quod de eo possit exire.'

[86] *Ann. Vet. Mut.*, col. 60.

[87] Remigio de' Girolami, *De Bono Pacis*, ed. Maria Consiglia De Matteis, in *La 'teologia politica comunale' di Remigio de' Girolami*, (Il mondo medievale, 3; Bologna, 1977), 123–4.

Since Gerard continued to envision the ban as the principal weapon against peace-breakers, he devoted several new statutes to rendering its use more equitable and systematic. Those citizens whose names he removed from the list of those under the ban for peace-breaking, because of reconciliation or relaxation of punishment, were specifically guaranteed 'all civil rights and benefits' enjoyed by other citizens.[88] Beyond this protection of rights, Gerard absolved from the ban all those banned for any reason other than peace-breaking, murder, or perjury, for which crimes he reserved to himself the power to negotiate peace accords.[89] Gerard declared his procedures for amnesty and arbitration to be normative for settling disputes in the city of Parma.[90]

Although there is no way of knowing how long and how well Gerard's— or any other Alleluia preacher's—reforms to ensure peace survived, the practical steps that he took to prevent recurrence of violence left their mark on the city's laws. His general programme is clear. First, he sought, by making the inflicting of punishment a matter of public record, to place it above the manipulations of individuals and factions. Second, by relaxing the most severe punishments and creating a method for absolving those guilty of crimes, he allowed for a 'safety-valve' in the justice system. Finally, by a large-scale amnesty, he allowed a 'clearing of the air' and opened the way to more lasting reconciliation and concord.

On the other hand, nothing in the friar's programme seems exceptional in comparison to the justice systems of the two other cities for which contemporary evidence survives, Vercelli and Verona. The effect of Gerard's activity was not the creation of some new means for preventing violence but the reconciliation of actual disputes within the city and between the city and other entities. Mediation, not legislation, occupied most of the friars' time. Their legislation itself never strayed far from the reconciliation of actual disputes or their causes, even when it treated such seemingly unrelated matters as usury or women's dress.

So, in conclusion, the friars' preaching, mediation, and legislation were aspects of the same programme. They helped create the concord and unity that characterized that 'time of tranquillity and peace' remembered by its participants as the Great Devotion of 1233.

---

[88] Parma Stat., 306: 'omnia . . . jura et beneficia civilia'.
[89] Parma Stat., 312; for use of proxies, ibid. 313.
[90] Parma Stat., 303–4.

Conclusion
Revivals and Politics

THE Great Devotion lasted a mere 10 months, from February to November of 1233. Its peak, during which John conducted his great campaigns and Gerard reformed the statutes of Parma, was even shorter, lasting from April to September. A seventeenth-century Vicenzan chronicler, Giambattista Pagliarini, convinced that his sources had erred, spread the events of John's campaigns out over 3 years.[1] It was an understandable error. The remarkable amount accomplished in so little time remains, even today, something of a marvel. Perhaps only the rapidity with which the revival ended is more surprising. Now that the character of the revival is clear, one needs to ask how a movement of such magnitude and power could collapse so quickly. As the preacher whose fame was greatest and whose fall was the most dramatic, John of Vicenza best illuminates the developments that brought the Great Devotion to grief. The Vicenzan Gerardo Maurisio said that John was brought down by his own lust for power and high political office.[2] Most recently, André Vauchez has suggested that the revival collapsed because the friars lacked the political power to carry out their programmes.[3] The only historian to attempt an overall history of the Devotion, Carl Sutter, proposed other reasons for John's downfall—in particular, a negative reaction to his burning of heretics at Verona and his naïvety in trusting Ezzelino. For Sutter, the pacification of the Veneto was impossible.[4]

In retrospect, one might wonder if any of these analyses are correct. It is tempting to believe, that John destroyed himself by creating unreasonable expectations in those to whom he preached. His followers turned

---

[1] Giambattista Pagliarini, *Cronaca di Vicenza*, ed. and trans. G. Alcaini (Vicenza, 1663), 37; id., *Chronica Vicentiae*, Vicenza, Biblioteca Maciana MS CCCLXVI, fo. 42ᵛ. And following him, Basilio da Schio, *Biografia del beato Giovanni da Schio*, Vicenza, Biblioteca Comunale MSS E. J. 13, fos. 376ᵛ–377ᵛ.

[2] Maurisio, 34.

[3] Vauchez, 546.

[4] So Sutter, 141–2; likewise, D. A. Brown, 'The Alleluia: A Thirteenth Century Peace Movement', *Archivum Franciscanum Historicum*, 81 (1988), 16, who suggests that papal–imperial rivalry and the 'absolute perversion of Ezzelino' made peace impossible.

against him because he could not deliver the impossible: a world of perfect concord and peace. Perhaps this is true, but even so, it alone cannot explain John's fall. In Bologna he miscalculated in his arbitration and then revised his decree without fatal consequences. He could still modify his programme and meet the demands of his hearers. He met his defeat in the Veneto so quickly that his hearers did not even have time to see if his peace plan would work. That John was vain and arrogant none would deny, but his craving for power was not itself his undoing. Rather, his chief errors were those of judgement. He forgot that the peace-maker could not dictate terms but could only bring into the open the unspoken or unspeakable desires of his hearers.[5] John could act high-handedly and even rashly—as in his burning of the Veronese heretics. This was not a fatal blunder: the citizens of Verona remained his loyal supporters until the very end. He had not contradicted their deepest wishes. In at least two cases, however, John did ignore the deep wishes of the people. He failed to appoint an impartial *podestà* in Vicenza, and he insisted on handing what amounted to a complete victory to the da Romano faction in the arbitration for Conegliano. One act undermined his support in Vicenza; the other gave his Paduan enemies a rallying-point.

Yet these mistakes need not have destroyed him. Unfortunately, he had compounded them by creating the impression that he consistently favoured the da Romano over the Lombards.[6] His decrees would have allowed the da Romano to consolidate their control in Treviso and Verona and forced reception of the exiled pro-da Romano faction into the pro-

---

[5] On this limitation on arbiters, see G. Kingsley Garbett, 'Spirit Mediums as Mediators in Korekore', in John Beattie *et al.* (ed.) *Spirit Mediumship and Society in Africa* (New York, 1969), 119.

[6] On John's favouritism and his loss of Paduan support, see Antonio Rigon, 'Chiesa e vita religiosa a Padova nel duecento', in *S. Antonio 1231–1981: Il suo tempo, il suo colto, e la sua città* (Padua, 1981), 292. Alfred Hessel calls this John's principal error, *Storia della città di Bologna dal 1116 al 1280*, ed. and trans. Gina Fasoli (Bologna, 1975), 108. Rigon, 'Vescovi e ordini religiosi a Padova nel primo duecento', in *Storia e cultura a Padova nell'età di Sant'Antonio* (Fonti e ricerche di storia ecclesiastica padovana, 16; Padua, 1985), 149–50, suggests that John's personal political power in Verona and Vicenza had grown too great not to alienate someone.

Why did John so markedly favour the da Romano? Basilio da Schio, *Monumenta de Iohanni Schio Vicentino ex Historicis Res Antiquas Marchiae Travisinae Scribentis*, Rome, Archivum Generale Ordinis Praedicatorum MS Fondo Libri HHH, fo. 377[r], says that some attributed his favouritism to family connections: '[D]alla suddetta prigionia del beato Giovanni presero occasione alcuni scrittori d'interpretar sinistramente le sue sante operationi, attribuendo tutto al sangue, alla carne, et ambitione di essaltar li amici e parenti, trà quali erano in quel tempo due valorosi capitani Martio e Matteo Schii favoriti di Fedrigo Secondo.' That John was related to the 'valorosi capitani' is doubtful, but that family considerations influenced John's actions is not implausible.

Lombard cities. In return, the Lombards gained little except the da Romano's cession of their Paduan holdings, and for the return of these lands Padua paid an exorbitant indemnity. The Paquara peace-terms gave John's enemies a valid grievance. This grievance could then be skilfully exploited by Jordan Forzaté, who, out of spite and fear, orchestrated the conspiracy that destroyed the friar and the peace. When it collapsed, John's peace did not simply fall apart from internal weaknesses; it had already been systematically undermined from without.[7]

Seen as a whole, John's decline was more spiritual than political.[8] His career suggests an explanation of why the Devotion passed so quickly. After his political reverses came something far more serious. As he lost control, John turned more and more to military force. Doing so, he became, little by little, no longer the Great Prophet of God, but another power-broker on the Veneto scene. His imprisonment broke the magical spell; his charisma departed, he was no longer invincible. The persona of the agent of God that he had so skilfully created by preaching and miracle-working vanished, never to return. Perhaps it is an unavoidable recipe for disaster when leaders whose authority is spiritual in origin take on political roles. They seem slowly but surely to assimilate themselves to the other participants in the political arena. That done, the reputation for impartiality and divine blessing which allowed their rise to power inevitably dissipates. The revivalist becomes a politician like any other— a role for which his style is maladapted—or he simply fades away. John seems to have tried the first alternative, only to undergo the second.

The obscurity of his later life is in vivid contrast to his glory. After September of 1233, John returns to the darkness from which he had suddenly emerged in April. He seems to have preached in the Veneto against Frederick II in 1236, since at that time the emperor wrote an angry letter to Pope Gregory IX protesting against John's activities. The pope, rather disingenuously, replied that he knew nothing of the preacher and had no control over his actions.[9] Gregory may have lost faith in the

---

[7] In one sense, it was only a temporary set-back for the plan. When the Veneto was finally pacified in 1236 it was largely along John's terms of 1233, as Sutter, 166–7, notes. And, in the end, this facilitated Ezzelino's domination of the region, just as the Paduans had feared.

[8] Here I would disagree with the recent opinion of D. A. Brown, 'The Alleluia', 5, who sees John's major weakness as a lack of '*political* acumen'.

[9] See Frederick's letter of protest in H.-B., III:II, 907–08. For the pope's reply, see Rodenberg, i. 702. On this crusade against Ezzelino the best summary remains Edouard Jordan, *Les Origines de la domination angevine en Italie* (Paris, 1909), 77–8; for the wider phenomenon of political crusades, see N. J. Housley, 'Politics and Heresy in Italy: Anti-Heretical Crusades, Orders and Confraternities', *Journal of Ecclesiastical History*, 33 (1982), 193–8.

Prophet of Bologna, but his successor, Innocent IV, seems to have been of a different mind. On 13 June 1247, the pope directed a letter to John as inquisitor of Lombardy.[10] In this letter he encouraged him to use care in reconciling heretics and granted an indulgence of twenty days to those who attended his sermons. Nevertheless, there is something ominous about this letter. While supporting John, the pope also reprimanded his Dominican superiors—he ordered that they not hinder John's activity and he reminded them that, if they failed to give John support, he had given the friar the authority to select any Dominican he wanted to be his companion.[11] Was this the occasion when, as Salimbene reports, John announced to the frustrated Dominican brothers who criticized him, 'It was I who exalted that Dominic of yours, who had lain hidden in the earth for 12 years, and if you do not keep quiet, I shall destroy your saint and publish your doings to the world'?[12] John's commission as inquisitor of Lombardy may have lasted as long as 4 years. It was probably over by 1251, when the pope appointed Peter of Verona to the office of inquisitor for Milan and its vicinity.[13] Peter's own career as inquisitor was brief. The heretics murdered him within 9 months.

After his appointment as inquisitor John disappears for many years. There is only one hint that he remained active in north Italian politics after the 1240s. Rolandino of Padua reports that a 'venerable brother John of the Order of Preachers' was with the pro-papal army under the pope's legate, Bishop Philip of Ravenna, when it fought the da Romano near Padua in 1257.[14] It is not impossible that this venerable brother John is some other Dominican. If he is John of Vicenza, this would imply a remarkable reversal in his known policy of favouring the da Romano. About 2 years later, John again reportedly received a vision, this one

---

[10] *BOP*, i. 174, letter 178.

[11] Innocent IV in a letter of 1251, Rodenberg, iii. 96–7, mentions a 'frater Johannes de ordine Predicatorum' as representing the commune of Naples at the Curia; if this was our John, he still enjoyed the pope's trust at that time. Savioli, i. 292, rejects this identification; later authors tend to accept it: Jacques Quétif and Jacques Échard, *Scriptores Ordinis Praedicatorum* (2 vols.; Paris, 1719), i. 153; Verci, ii. 210; Antonio Magrini, *Notizie di fra Giovanni da Schio per le nozze di Nanne Gozzadini e Maria Teresa Sarego* (Padua, 1841), 40. Sutter, 157, is uncertain. To be rejected is the identification of John as the papal agent who lifted the excommunication of Vicenza in 1257. That John was John of Vercelli.

[12] Salimbene, 79.

[13] His commission may not have been as extensive as John's, see Antoine Dondaine, 'Saint-Pierre-Martyr, Études', *AFP* 23 (1953), 100–1.

[14] Rolandino, 129 (for 'frater Johannes' in 1257). Since Rolandino usually gives John the title 'of Vicenza' or 'of Bologna' this identification is doubtful. Savioli, i, 292, rejects it. Verci, ii. 210 would make it. Sutter, 158, again is doubtful.

rather ambiguously predicting the coming death of his Dominican brother Ralph.[15] Some 3 years pass before the last news of John—which comes, rather appropriately, in a miracle-story. Leandro Alberti tells us that it was in 1262 that John publicly predicted the election of John of Vercelli as master of the Order of Preachers. The only thing known of him after that date is that he died in Apulia, probably in the 1260s.[16] John's glory had passed long before that time.[17]

The other preachers of the revival disappeared into equal or greater obscurity. Leo de' Valvassori and Guala of Bergamo, having lost their reputation as revivalists, served as bishops of Milan and Brescia respectively. In spite of that honour, they were forgotten outside the devotion of their religious orders. Gerard went on to a long career as a Franciscan preacher, but no one recorded many details of it. In the end he retired to his home town of Modena, where he died and reportedly appeared in a vision to Jacopino of Reggio. His Franciscan brethren buried him in an honourable stone tomb in their church. Jacopino's glory seems to have faded after 1233. Little evidence exists for his later life, apart from a few anecdotes in Salimbene. These show that he continued to enjoy a reputation as a visionary and a thaumaturge. After his death he was buried in a respectable tomb at the Dominican church in Mantua and promptly forgotten.[18] Skilled revivalists, then, can end their lives in relative obscurity, even after achieving considerable fame and glory.

The fragility of the revivalists' power was mirrored in their unstable connection to the ecclesiastical hierarchy. John of Vicenza's ambiguous relationship to the communes of Italy, the papacy, and his own order highlights the tension between the revivalists and the secular and ecclesiastical establishments. To the secular powers the preachers were always outsiders. They functioned as city managers principally because more than any of the others available they were above faction and party—they were pure, uncontaminated by political alliances. This independence provided remarkable room for manœuvering. John could issue a decree apparently favouring the bishop of Bologna over the

---

[15] *VF*, 275.

[16] Salimbene, 73. Francesco Barbarano, *Historia ecclesiastica della città, territorio e diocesi di Vicenza* (Vicenza, 1649), 95, would have him take part in the crusade against Manfred in Apulia during 1265. This is highly unlikely since Th. Cant., 424, speaks of him as already dead in 1263. Perhaps John went to Apulia in 1259 to preach against Manfred after his excommunication; if so, Barbarano's story has some factual basis. So Sutter, 160.

[17] On his late and ephemeral cult in Vicenza, see Sutter, 160–3.

[18] Salimbene, 73.

commune, then reverse it, and suffer no fatal consequences. On the ecclesiastical front, there is no indication of how the local bishops received the revivalists, excepting John's special relation to his patron, the bishop of Modena. Bologna's bishop, certainly, did not find in John a convenient supporter of episcopal rights—John seems to have favoured the commune.[19] Later, in the Veneto, when his arbitrations had the impact of favouring the imperial party, they won him the enmity of pro-Lombard ecclesiastics like Jordan Forzaté, the spiritual father of Padua. It is evident that unpredictable mendicant preachers, who escaped episcopal jurisdiction, received at best a cautious welcome from most resident bishops. The bishops' desires, however, mattered little. The revivalists' authorization to preach came from secular communal governments and, occasionally, through papal letters.

As mendicant friars, the preachers were also subject to the superiors of their own order and, above all, to the pope. Nevertheless, a preacher's popularity flowed from his charismatic appeal, not from the approbation of Church authorities. When a religious order like the Dominicans legislated about its preachers, it could do little more than provide rudimentary training and cull out the obviously inept or politically dangerous.[20] John of Vicenza, at least, seems to have felt cramped by his subordination to his order. When the master of the Dominican Order, Jordan of Saxony, attempted to get his subject to move on from Bologna during the spring general chapter of 1233, it came to nothing.[21] Jordan was thwarted by the will of the commune and John's unwillingness to co-operate. Jordan could have disciplined the independent-minded friar, but only at the cost of alienating most of Bologna and, one suspects, a good number of the local Dominicans. It is striking that Jordan's desire to make John move on may well have resulted from his knowledge of Pope Gregory's repeated requests that John come to preach in Tuscany.[22] When the time to move finally came, John, defying both master general and pope, went not south but north to the Veneto.

Perhaps the preachers' most visible link to an ecclesiastical authority was that with the papacy. From the papal side, support seems almost wholly opportunistic; from the preacher's side, it seems inconsequential. A papal mandate did not create a charismatic preacher, and its reception

---

[19] For his final judgement, favouring the commune, see Savioli, iii. 132.

[20] Legislation in *ACPR*, 5, 11 and *ACPL*, 145.

[21] As in *VF*, 138–9; repeated by Borselli, *Cron.*, 22–3.

[22] For the pope's letters to the Dominican chapter and the commune, see *BOP*, i. 48–9, letters 74 and 75 (Auvray, i. 713–14, letters 1268 and 1271).

by a proven preacher added little to his appeal. Peter Martyr was not above preaching without approval, and the lack of it never hurt the size of his crowds. When a pope like Gregory IX did become involved in a revival, he invariably seems to have arrived late and out of breath. Gregory wrote John flattering letters,[23] cajoled him to become involved in papal projects,[24] and ratified his sermons with grants of indulgences.[25] John showed little interest in Gregory save as another name to add to the list of those authorizing him as an arbiter, which he regularly included in his decrees. But for the pope, John was a useful commodity, immensely popular and apparently orthodox. An astute pope would enlist him in whatever the current papal cause, if he could. Yet, if the history of Gregory's solicitation of John is typical, a pope's chances of co-opting a successful preacher were slim, and the pope remained ready to abandon a troublesome preacher when that was convenient. Ultimately, papal approbation followed a preacher's success and did little to enhance his reputation or control his actions, and the preacher probably knew this.

For Salimbene and others of his time, the Devotion of 1233 was the most remarkable event of the age. Why?

First, because of the intensity of the popular response to the preachers. The revivalists worked as a group; they trumpeted each other's successes, and they promoted each other's activities. The preachers established a rapport with their hearers that allowed them to sense the crowd's hopes and fears, speak to them, and finally move those hearers to action.

Secondly, not content with merely spiritual renewal or individual moral reform, the preachers took an active role in reconciling conflicts between individuals, families, corporations, and even cities. This massive wave of peace-making, coming after decades of war within and between the communes of the region, appeared to many a direct intervention of God.

Thirdly, as the revival progressed, it took on a dimension that is, for its scope and character, unique in the Italian middle ages. In one city after another, control of the municipal government passed to the preachers, who revised and corrected the city's laws and statutes, introducing new legislation on their own initiative. Then, as peacefully as they had taken power, they relinquished it to depart and carry the revival to

---

[23] *BOP*, i. 51, letter 78 (Auvray, i. 751, letter 1339).
[24] *BOP*, i. 48, letter 73 (Auvray, i. 713, letter 1270).
[25] *BOP*, i. 57, letter 88 (Auvray, i. 813–14, letter 1461).

other cities. While in later centuries there were many popular, even revivalistic, preachers—Bernardino of Siena and Savonarola, for example— the phenomenon of a group of preachers conducting their revival as a group and then systematically legislating for the cities in which they preached is unparalleled.

The revivalism that the friars of the Devotion practised had a longer life than the revival itself. Preachers like the Dominican Theobald of Albinga and the Franciscan Gratian of Padua already practised an Alleluia-like revivalism in the 1220s. Preachers in the same style, like Ambrose Sansedoni and Peter of Verona, continued to employ similar techniques well into the second half of the century. If the Alleluia itself was short-lived, certain aspects of its religiosity did not go out of fashion for several decades. Still, these later practitioners failed to spark any new 'devotions'. Finally, in the 1280s, Salimbene could speak of the 'the way of the ancient preachers' as something of the past.

What explains the great outburst of the Devotion itself? It could be explained as the result of spontaneous emotional contagion. This has been a popular interpretation.[26] To me it seems unconvincing. The enthusiasm of the revival did not occur spontaneously; it had to be elicited, cultivated, and carefully nurtured. The Great Devotion was highly emotional, but the friars excited such fervour principally because they identified and spoke to their hearers' deepest needs and desires: the need for debt relief, the yearning for an end to senseless bloodshed, and the hunger for divinely guided justice. It was these needs that made the revival possible. Nevertheless, contrary to a mechanistic political or economic explanation, they did not make it inevitable. Had the friars not found the proper religious vehicle for appealing to their hearers, they might never have provoked such an enthusiastic response, and, as a result, there would have been no great revival. As a concrete example of the ease of failure, one has only to recall how the mob ran an otherwise

---

[26] At the turn of the century that was the opinion of G. G. Coulton, *From St. Francis to Dante: Translations from the Chronicle of the Franciscan Salimbene (1221–1288) with Notes and Illustrations from other Medieval Sources*, 2nd edn. (1907; repr. Philadelphia, 1972), 21, following J. A. Symonds, 'Religious Revivals in Medieval Italy', *The Cornhill Magazine*, 31 (1875), 54, who explained such revivals as the result of the 'Italian temperament' and their lack of 'any great depth of moral earnestness'. Amazingly enough, such ideas are not yet out of style: Giulia Barone, 'L'Ordine dei Predicatori e le città: Teologia e politica nel pensiero e nell'azione dei predicatori', *MEFR* .89 (1977), 615, sees the Alleluia as based on an 'onda dell'emozione popolare'. Raoul Manselli, *La Religion populaire au moyen âge: Problèmes de méthode et d'histoire* (Montreal, 1975), 112, similarly sees the Alleluia (and the later Devotion of 1260) as essentially spontaneous, the leaders being simply a point of departure.

successful preacher, Roland of Cremona, out of Piacenza after he preached an opening homily on the suppression of heresy.[27] He had failed to touch the right nerve.

William G. McLoughlin, in his very perceptive study of modern revivalism in the United States, has written, 'Since revivalists alone cannot manufacture revivals and since social crises do not automatically produce them, their explanation must lie in a particular combination of men and events.'[28] The particular response to the present crises by the revivalists involved determines the sort of revival produced. A union of three attributes distinguishes the Alleluia: miracle-working, peace-making, and statute reform. The miraculous element was introduced and fostered by the very preaching-style of the friars. When Anthony of Padua shunned miracle-working, that was sufficient to set him apart from the Alleluia style. Many others, from John of Vicenza to the present, have made miracle-working and divine revelations a part of their message. This aspect of the Devotion has a long history in revivalism. On the other hand, few revivals if any sparked campaigns of reconciliation and legal reform such as marked the Devotion of 1233.

For a revival like the Alleluia to take on the complexion of a peace movement, and then pass to legislative activity, required, beyond a certain style of preaching, a backdrop of particular social and political conditions: rampant factionalism and a growing sense that the violence provoked by it had become intolerable. That northern Italy was divided and factious in the 1220s and 1230s is indisputable.[29] The Veneto had not yet been forcibly pacified, as it would be after 1237 under the tyranny of Ezzelino da Romano.[30] Emilia and the Romagna would remain in a

---

[27] Mussi, col. 461.

[28] William G. McLoughlin, jr., *Modern Revivalism: Charles Grandison Finney to Billy Graham* (New York, 1959), 7. Gatti, 'Religiosità populare e movimento di pace nel'Emilia del secolo XIII', in *Itinerari storici: Il medioevo in Emilia* (Carpi, 1980), 84–5, emphasizes the elements of crisis that made north Italy ripe for a revival in 1233. She lists conflict of Church and state, social unrest, bad weather, and fear of portents and diabolical forces. This background is not enough to explain the events of 1233. Gatti's examples are mostly from the 1220s, and the 1230s do not seem more 'in crisis' than any other decade of the 13th cent.

[29] On the impotency of the empire in northern Italy during our period and the political fragmentation of the region, see Paolo Brezzi, 'I comuni cittadini italiani e l'Impero medioevale', in *Nuove questioni di storia medioevale* (Milan, 1969), 190–4.

[30] Ezzelino and his party would go on to become the pillars of the imperialist party in north Italy during the mid-1200s. See J. K. Hyde, *Society and Politics in Medieval Italy: The Evolution of the Civic Life, 1000–1350* (New Studies in Medieval History; London, 1973). 122.

state of division and unrest for years to come.[31] The desire for peace-making could arise when the citizens of an area became disgusted by continuous unproductive warfare or, as seems to have happened in northern Italy during the early 1230s, economic crisis dictated an end to hostilities. Exhaustion and war-weariness were not unique to the time of the Alleluia. Similar sentiments doubtless underlay the peace-making by Ambrose Sansedoni at Siena during the 1250s and that by Peter of Verona in the Romagna during the 1240s. Peter's activity may even be linked in some way to the Great Devotion. Bologna itself entered a new period of civil strife in the early 1240s. Temporary relief came with the election of a new bishop in 1244—none other than John's protégé, James Boncambio, whose conversion to the religious life in 1233 had enlivened one of the revivalist's sermons. The municipal chronicle records laconically that many feuds were finally ended and peace established in that year.[32]

So if there was a desire for peace-making and reconciliation and the revivalist addressed that need, he could add that element to his revival. Nevertheless, later peace-campaigns by preachers never took on the magnitude of that of 1233, and none became a full-blown movement for statutory reform in which the revivalists took direct control of city governments. A preacher like Savonarola, powerful as his influence was in city politics, never held public office. For a revival to enter a stage of statutory reform required more than a simple desire for peace. Ambrose Sansedoni encouraged laws to defend widows and orphans at Siena, and Henry of Cominciano legislated against heresy at Vercelli. Still, when compared with Gerard's legislation at Parma, these were *ad hoc* reforms. The widespread installation of preachers as city managers with broad arbitrary powers is distinctive of the Alleluia alone.

To explain why the friars became the city managers of choice in 1233 is not easy. One precondition that suggested calling in such outsiders and granting them extraordinary legislative power may have been the rampant instability and consequent lack of legitimacy in the region's governments during the period. When the forceful entry of previously disenfranchised groups destabilized a communal government, and no universally accepted means to redistribute power could be found, the citizens could choose to hand over the city government as a whole to a trusted moderator. Virtually

---

[31] I have described these conditions in Ch. 6.

[32] CCB, 119–20; Rampona, 119: '1244—In questo anno si fece la pace di tutte le guerre che erano nella città di Bologna.'

every city in the Veneto underwent a painful expansion of the franchise during the 1220s and 1230s.[33] The entry of the *popolo* into the communal government was a gradual process, involving the creation of new corporate entities like the Bolognese Società delle Armi and Società delle Arti. Under such unstable conditions, the friars, with their aura of holiness and divine authority, must have appeared as an ideal choice for civic moderator. They were men who could introduce stability and act as arbitrators and legitimizers of new arrangements in communal government.

Does this explain the preachers' role as city managers? One good argument against this hypothesis exists: in the friars' statute reforms there survives not a single piece of 'constitutional' legislation. On the other hand, such legislation was unnecessary. The legitimization of a new order of government could be achieved by simply removing those officials who refused to accept it. The Vicenzans, one remembers, invited friar John to his home town 'to remove the *podestà* and appoint one who would be acceptable to all'.[34] Whether legitimacy required a change of laws or a change of personnel, the revivalists of 1233 seemed the right men for the job. By the end of the decade this particular type of political instability began to disappear from the region, and with it the widespread use of friars as city managers.[35] Preachers continued to preach and put themselves forward as workers of miracles, but they no longer acquired political powers like those of 1233.

Contemporaries of the Alleluia, like Gerardo Maurisio, and modern

---

[33] See Gina Fasoli, 'Oligarchia e ceti popolari nelle città padane fra il XIII e il XIV secolo', in Reinhard Elze and Gina Fasoli (ed.), *Aristocrazia cittadina e ceti popolari nel tardo medioevo in Italia e Germania* (Annali dell'Istituto Storico Italo-Germanico, 13; Bologna, 1984), 11–40. Some examples: at Modena expansion of the franchise began in 1229, ibid. 25; at Parma the organization of the Arti occurred in 1211, but they struggled for a recognized role in the commune until 1244, when they entered the Consiglio Cittadino, ibid. 27; Piacenza suffered from continuous civil war after 1218 over the roles of the *popolani* and the *nobili* in the government, ibid. 28; although the populars had entered the government of Vicenza in 1222, their role in government remained contested until the domination by Ezzelino in 1237, ibid. 35; in Verona, beginning in 1227, the populars struggled with the *milites* for a role in the government until their advance was checked by Ezzelino, Andrea Castagnetti, 'Appunti per una storia sociale e politica delle città della Marca veronese-trevigiana (secoli XI–XIV)', in Reinhard Elze and Gina Fasoli (ed.), *Aristocrazia cittadina e ceti popolari nel tardo medioevo in Italia e Germania* (Annali dell'Istituto Storico Italo-Germanico, 13; Bologna, 1984), 58–9. The one city where the populars were completely excluded from power, Ferrara, experienced no revival in 1233; see Fasoli, 'Oligarchia e ceti popolari', 25.

[34] Maurisio, 32.

[35] I find one notable exception: when Bologna underwent a reorganization of government in 1262, they called on two Frati Gaudenti to reform and codify their statutes.

scholars, from Sutter to Vauchez, agree that the revival and its peace-making failed 'and was considered as nothing'.[36] Is this completely correct? If one reckons the success of a revival by its ability to create a permanent state of religious exaltation, then the Alleluia, like all revivals, was a failure. Enthusiasm eventually wanes. For most observers, the failure of the Devotion is identical with the collapse of John's peace programme for the Veneto. Paquara was a failure. Nevertheless, that failure should not obscure the many reconciliations that did endure. The communes continued to reaffirm them decades later. Likewise, the friars' legal reforms provided lasting means to protect the weak and to arbitrate disputes. These laws remained on the books, at least in Parma. If considered on a less ambitious scale, then, the preachers' programmes succeeded. These secular effects were the most enduring product of the revival. Measured in terms of that achievement, the Alleluia was not a failure, even though it did not bring universal concord to northern Italy.

The Devotion's effect on the religious life of the communes seems much less substantial. The friars do seem to have harnessed popular fervour for the construction of new churches and convents. The Devotion inspired construction at Reggio, Parma, and Modena.[37] Dominican foundations during 1233 at Como and Reggio might be linked to the preaching of John of Vicenza.[38] If the enthusiasm of the Great Devotion soon vanished, it did leave a physical mark on some cities in which the friars preached.[39]

Vito Fumagalli, among others, has suggested that the friars insti-tutionalized the revival by channelling the religious enthusiasm of 1233

---

[36] Maurisio, 33–4: 'pro nichillo reputatum est.'

[37] For Reggio, see Alberto Milioli, *Liber de Temporibus et Aetatibus et Cronica Imperatorum*, ed. Oswald Holder-Egger (MGH.Ss. 31:509); Parma, *Chronicon Parmense ab Anno 1038 usque ad Annum 1338*, ed. Giuliano Bonazzi, (RIS²9:9; Città di Castello, 1902), 10, and Salimbene, 103; Modena, *Ann. Vet. Mut.*, col. 60. The suggestion of Vito Fumagalli, 266, that the roofing of the Cathedral of Bologna in 1233 was a result of the Alleluia, is not convincing—Villola, 103, specifically ascribes its initiation to one Magister Tura.

[38] Vladimir J. Koudelka, 'La fondazione del convento domenicano di Como (1233–1240)', *AFP* 36 (1966), 397–8, suggests John's involvement with Como; see Stefano L. Forte, 'Le province domenicane in Italia nel 1650', *AFP* 41 (1971), 427, on the connection with Reggio. Forte's source is the unreliable Piò, *Vite degli huomini illustri di S. Domenico* (Bologna, 1620), i. 64. The foundation in Reggio was the work of Jordan of Saxony; see Marguerite Aron, *St Dominic's Successor* (London, 1955), 170.

[39] Gary Dickson, 'The Flagellants of 1260 and the Crusades', *Journal of Medieval History*, 15 (1989), 259, n. 4, has noted the similarity of this building programme to that during the penitential activity of the Nudi in Normandy during 1145. On the Nudi, see L. Delisle, 'Lettre de l'abbé Haimon sur la construction de l'église de Saint-Pierre-sur-Dive en 1145', *Bibliothèque de l'École des Chartes*, 21 (1860), 113–39.

into the creation of lay confraternities, some of which were dedicated to fighting heresy.[40] The evidence for such a contention is weak. No concrete evidence exists that any Franciscan confraternity or order of penance resulted from the Alleluia.[41] On the Dominican side, Peter of Verona did found a 'Confraternity of the Faith' at Milan in 1232, and Bartholomew of Breganza organized the 'Militia of Jesus Christ' at Parma during the year of the Great Devotion itself.[42] Linkage of these foundations with the Alleluia is based on the perception of the revival as in great part an anti-heresy movement, a position recently presented with vigour by Gary Dickson.[43] Showing that Alleluia preachers directed campaigns against heretics is different from proving that the Devotion of 1233 was predominantly that sort of campaign. Bartholomew was a preacher of 1233 and Peter did preach in the 'Ancient Style'; both organized confraternities to combat heresy. But the linkage of these confraternities with the Alleluia is tenuous. Peter's confraternity was part of the long Dominican tradition of organizing lay committees to report on heresy. Such committees originated during Dominican preaching in Languedoc, perhaps as early as 1217.[44] But Peter, as far as we know, was never directly involved in the revival, and his committee was founded a year before it occurred. Bartholomew's Militia can claim an Alleluia preacher as its founder and the year of the Devotion as its date of establishment, but that is all. The Militia of Jesus Christ consisted of a select number of nobles, and its constitutions drew their inspiration from those of the Cistercian-inspired military orders of the previous century.[45] Such an élite group could never

---

[40] Fumagalli, 268.

[41] Contrary to the unsubstantiated suggestion of Felice da Mareto, 'L'ordine francescano della penitenza a Parma, Fidenza, Piacenza e Modena', in Mariano d'Alatri (ed.), *Il movimento francescano della penitenza nella società medioevale: Atti del Terzo Convegno di Studi Francescani, Padova, 25—26—27 settembre 1979* (Rome, 1980), 312, that the confraternity of penitents of S. Francesco Piccolo at Parma arose during the Devotion, there is no evidence of their existing before 1238—as Mareto himself concedes.

[42] See Gilles-Gérard Meersseman, 'Les Milices de Jésus-Christ', *AFP* 23 (1953), 275–308, on these groups.

[43] See Dickson, 'The Flagellants of 1260', 247, where he explicitly affirms the similar positions of Vauchez and Housley.

[44] Meersseman, 'Les Confréries de Saint-Pierre-Martyr', *AFP* 21 (1951), 276–7.

[45] So ibid. 305. If only because this mistake continues to be made—e.g. Vicenzo Petriccione, 'I penitenti francescani di Bologna nel secolo XIII', in Mariano d'Alatri (ed.), *I frati penitenti di san Francesco nella società del due- e trecento* (Secondo Convegno di Studi Francescani, Roma, 12—13—14 ottobre; Rome, 1977) 262—it must be repeated that the Militia was in no way connected to Bartholomew's other creation, the Frati Gaudenti of 1261; see Antonio de Stefano, 'Le origini dei Frati Gaudenti', in *Riformatori ed eretici del medioevo* (Palermo, 1938), 221–4. For the Rule of the Militia of Jesus Christ (1233), see

have served to channel popular religious enthusiasm.[46]

This is not to say that the friars' preaching was without effect, perhaps an indirect one, on heretics. Stephen of Spain testified at the inquest for Dominic's canonization that the revival brought about the conversion of 'one hundred thousand heretics to the Catholic Faith'.[47] There is no way to confirm this testimony or determine how permanent these individual conversions were. Likewise, it will never be known how many orthodox hearts were changed or lives reformed by the preaching of the friars. Such spiritual fruits lie forever hidden from the historian. Beyond this point the historian cannot venture in the search for the effects of the revival.

If one seeks the enduring legacy of the preachers of 1233, it was not a set of particular benefits to the people of northern Italy—their peace-making, their legislation, the building-projects that they inspired, or even the hearts they moved. The real legacy was the new type of relation between preacher and audience that the friars excelled at creating.[48] An ability to sense the needs and fears of one's hearers, to give them expression, and to suggest concrete actions to resolve them would become the mark of all successful preachers in the centuries that followed. The Alleluia preachers did not invent this special form of contact between preacher and audience, but they did show its power. So if later fam-ous preachers, like Berthold of Regensburg, Bernardino of Siena, or Savonarola, sometimes resemble the more obscure friars of 1233, this is because they too achieved that special relationship with their hearers that was the essence of the 'way of preaching of the ancient preachers'.

Dossier de l'ordre de la pénitence au XIIIe siècle, ed. Gilles-Gérard Meersseman (Spicilegium Friburgense, 7; Fribourg, 1961), 290–5. Those who would like an overview of confraternities, including their role in peace-making after the period under consideration, should consult Gilles-Gérard Meersseman, Ordo Fraternitatis: *Confraternitate e pietà dei laici nel medioevo* (3 vols.; Italia Sacra, 24–6 Rome, 1977), which incorporates his earlier works to which I refer. A good sense of the contemporary field of confraternity studies may be had in *Le Mouvement confraternel au moyen âge* (Collection de l'École Française de Rome, 97; Geneva, 1987).

[46] It could, of course, act as a guarantor for reconciliations made during the revival.

[47] *Act. Can. Dom.*, 158.

[48] So Carlo Delcorno, 'Origini della predicazione francescana', in *Francesco d'Assisi e francescanesimo dal 1216 al 1226* (Atti del IV Convegno Internazionale, Assisi, 15–17 ottobre 1976; Assisi, 1977), 160.

BORSELLI'S UNEDITED ACCOUNT OF 1233

What follows is the entry for the year 1233 in the *Chronica Magistrorum Generalium Ordinis Fratrum Predicatorum* of Girolamo de' Borselli.[1] It excludes, however, the canonization testimony for Saint Dominic, which seems to be digested from depositions in the Bologna convent. Although the text is late and was left unfinished at his death in 1497, Borselli, as resident in the Dominican convent of Bologna, had access to Dominican oral traditions and the Bolognese Dominican archives. His witness is especially important. The passage is in part parallel to that in his *De Rebus Gestis Pertinentibus Bononiae*, which depends directly on the Rampona Chronicle and the *Vitae Fratrum*. It adds, however, unique material, in particular John's otherwise unrecorded visiting and healing of the sick.

I have preserved the spelling of the original, but modernized the capitalization and the punctuation. The paragraph numbers in the margin are from the original.

Text: Bologna, Biblioteca Universitaria MS 1999, fo. 21ʳ.

3 Anno Domini 1233, celebratum est quartum decimum capitulum generale Bononie sub magistro Iordane, in quo interfuerunt plusquam trecenti fratres. Frater Nicolaus, uir optime sanctitatis, in quodam loco orans tempore huius capituli et cogitans de sanctitate beati Dominici quia multi loquebantur de eius prima translatione, audiuit hanc uocem de beato Dominico: 'Hic accipiet benedictionem a Domino, etc.'ᵃ Tunc corpus beati Dominici nundum canonizati fuit translatum de sub terra in sepulcrum marmoreum satis grossum, in quo nulla erat ymago sculpta.

4 Hoc anno, frater Petrus de Verona, postea martir, adhuc iuuenis, fuit assignatus in conuentu mediolanesi.ᵇ

Hic, frater Iohannes Bononie predicans, cum usuras precipue detestaret, adeo populum inflamauit ut Landulphum usurarium publicum e ciuitate cum clamoribus exturbaret, dicendo: Pereat Landulphus et domus eius! Sicque furore populi domus eius exspoliata est et ipse lapid<at>us e ciuitate eiectus. Multas concordias inter ciues edidit. Mulierum pompas coercuit.

[1] Found in Bologna, Biblioteca Universitaria MS 1999. The text and its drawings of the masters of the order are in Borselli's own hand. The work is incomplete, more a scrap-book than a finished chronicle: many slips are inserted in the earlier sections and many pages are blank in the later sections. Since other recensions of the canonization process have been published—edited by Angelus Walz in MHD, 91–194, and in Échard, i. 44–56—I have only transcribed the part of the entry which pertains to John.

On Borselli himself, see Gianfranco Pasquali, 'Gerolamo Albertucci de' Borselli, O. P. (1432–1497): Ricerche bio-bibliografiche', *Rivista di storia della Chiesa in Italia*, 25 (1971), 59–82. Pasquali, 81–2, following modern opinion, considers Borselli a very trustworthy source because of his careful use of authentic ancient records.

Instituta urbis meliora fecit. Infirmos uisitans multos signo crucis liberauit et egrotationes incurabiles amouit. Docuit congredientes in nomine Yhesu Christi inuicem et mutuo salutare. Pridie ydus Madii urbem nudis pedibus tam religiosos quam laycos lustrare precepit, et deinde corpus beati Dominici humi collocatum in archa extolli fecit. Postea uero cum e Bononia discedere decernisset, aduocata populi contione, ciues ad concordiam, pacem, et honestatem fecundissime est hortatus, atque oranti, omnibus conspitientibus, crux in fronte ei apparuit. Que quidem res multos ad penitentiam redire admonuit. Tandem ad Galliam Cisalpinam est profectus, ubi multis miraculis claruit. Quo autem tempore, diem obierit compertum non habetur.[c]

(fo. 16[r]) Frater Jacobus de Boncanbiis bononiensis hiis temporibus hoc modo uenit ad ordinem. Cum enim famosissimus doctor esset Bononie, et propter ipsum multi scolares ex diuersis partibus uenirent ad studium, quodam die festo, equo albo insidens, indutus ueste pretiosa et torque aurea ad collum, casu uenit ad predicationem fratris Iohannis de Vincentia, qui predicabat in platea, et compunctus, cum illo ornatu uenit ad ecclesiam sancti Nicolai, admirante tota ciuitate, sancte religionis habitum suscepit. Hic fuit postea prouintialis Lombardie, uice cancellarius domini pape, et ultimo episcopus bononiensis.

5   (fo. 21[r]) Gregorius nonus tres legatos Bononiam misit circa hec tempora, qui inuestigarent de uita et miraculis beati Dominici ut eum inter sanctos connumeraret.[d]

---

[a] *Ps. 23: 5.*

[b] *Hinc usque ad* non haberetur *in fol. adfixo ad fol. 21[r].*

[c] *In margin.* Ponitur hic conuersio fratris Iacobi de Boncanbiis. Vide supra. *Cf. Borselli, Cron., 22.*

[d] *Sequitur processus canonizationis.*

BIBLIOGRAPHY OF WORKS CITED

MANUSCRIPTS

Bologna. Archivio di Stato MS: *Registro Grosso*.
—— Archivio di Stato MS: *Registro Nuovo*.
—— Biblioteca Universitaria MS 1794, fos. 70ᵛ–108ʳ: Bartholomew of Trent, [*Liber Miraculorum Beate Marie Virginis*].
—— Biblioteca Universitaria MS 1999: Girolamo Albertucci de' Borselli, *Chronica Magistrorum Generalium Ordinis Fratrum Predicatorum*.
Florence. Biblioteca Nazionale MS Conventi Soppressi, D. I. 937: Remigio de' Gerolami, *Sermones de Sanctis*.
Padua. Biblioteca Antoniana MS 466: Luke the Lector, OFM, *Sermones de Adventu et Festivis*.
—— Biblioteca Antoniana MS 477: anon., *Sermones de Tempore et de Sanctis*.
Rome. Archivum Generale Ordinis Praedicatorum MS Fondo Libri, HHH, fos. 370ʳ–388ᵛ: Basilio da Schio, *Monumenta de Iohanni Schio Vicentino ex Historicis Res Antiquas Marchiae Trevisinae Scribentium*.
Siena. Biblioteca Comunale MS T. IV. 7: Ambrose Sansedoni, *Sermones de Tempore*.
Vicenza. Biblioteca Comunale MS E. 2. 26: Lodovico da Schio, *Vita del b. Giovanni da Schio dell'Ordine de' Predicatori*.
—— Biblioteca Comunale MSS E. J. 12–15: Basilio da Schio, *Biografia del beato Giovanni da Schio* (4 vols.).
—— Biblioteca Maciana MS CCCLXVI: Giambattista Pagliarini, *Chronica Vicentiae*.

STATUTES

Bergamo. *Antiquae Collationes Statuti Veteris Civitatis Pergami*, ed. Giovanni Finarzi (Historiae Patriae Monumenta, 16: Leges Municipales, II; Turin, 1876).
Bologna. *Statuti di Bologna dall'anno 1245 all'anno 1267*, ed. Lodovico Frati (3 vols.; Bologna, 1869–77).
—— *Statuti delle Società del Popolo di Bologna*, i: *Società delle Armi*, ed. Augusto Gaudenzi (Fonti per la storia d'Italia: Statuti, secolo XIII; Rome, 1889).
Monza. *Memorie storiche di Monza e sua corte*, ii: *Codice diplomatico*, ed. Anton-Francesco Frisi (Milan, 1794).

Padua. *Statuti del comune di Padova dal secolo XII all'anno 1285*, ed. Andrea Gloria (Padua, 1872).

Parma. *Statuta Communis Parmae Digesta Anno 1255* (Monumenta Historica ad Provincias Parmensem et Placentinam Pertinentia, 1; Parma, 1856).

Treviso. *Gli statuti del comune di Treviso*, ii: *Statuti degli anni 1231—1233— 1260–63*, ed. Giuseppe Liberali (Monumenti storici pubblicati dalla Deputazione di Storia Patria per le Venezie, NS, 4; Venice, 1951).

Vercelli. *Statuta Communis Vercellarum ab Anno MCCXLI*, ed. Giovambatista Adriani (Historiae Patriae Monumenta, 16: Leges Municipales, II; Turin, 1876).

Verona (1228). *Liber Juris Civilis Urbis Veronae*, ed. Bartolomeo Campagnola (Verona, 1728).

—— (1276), *Gli statuti veronesi del 1276 colle correzioni e le aggiunte fino al 1323*, i. ed. Gino Sandri (Monumenti storici pubblicati dalla Deputazione di Storia Patria per le Venezie, NS, 13; Venice, 1940).

Vicenza. *Statuti del comune di Vicenza 1265*, ed. Fedele Lampertico (Monumenti storici publicati dalla Deputazione Veneta di Storia Patria, 2nd. ser.: Statuti; 4 vols.; Venice, 1886), i.

OTHER LEGAL SOURCES

AZZO DEI PORCI, *Summa*, (Venice, 1596).

Collectio Chartarum Pacis Privatae Medii Aevi ad Regionem Tusciae Pertinentium, ed. Gino Masi, (Orbis Romana, 16; Milan, 1943).

Corpus Iuris Canonici, ed. Aemilius Friedberg (2 vols.; Leipzig, 1922).

Corpus Iuris Civilis, 16th edn., ed. Theodor Mommsen (3 vols.; Berlin, 1954).

Formularium Florentinum Artis Notariae (1220–1242), ed. Gino Masi (Orbis Romanus, 17; Milan, 1943).

ODOFREDUS OF BOLOGNA, *Lectura in Digesto Veteri* (2 vols.; 1550; repr. Bologna, 1970?).

RAINERIUS PERUSINUS, *Ars Notariae*, ed. Ludwig Wahrmund (Quellen zur Geschichte des römisch-kanonistischen Processes im Mittelalter, 3; Innsbruck, 1917).

RICHARDUS ANGLICUS, *Summa de Ordine Iudiciario*, ed. Ludwig Wahrmund (Quellen zur Geschichte des römisch-kanonistischen Processes im Mittelalter, 2; Innsbruck, 1915).

ROLANDINUS PASSAGERII, *Summa Totius Artis Notariae* (2 vols.; Venice, 1588).

TANCRED OF BOLOGNA, *Ordo Iudicarius*, in *Libri de Iudiciorum Ordine*, ed. Friedrich Christian Bergmann (1842; repr. Göttingen, 1965).

PRINTED SOURCES

Acta Canonizationis Sancti Dominici, ed. Angelus Walz. (MOFPH 16: MHD 2; Rome, 1935), 91–194.

Acta Capitulorum Generalium Ordinis Praedicatorum, i, ed. Benedictus Maria Reichert (MOFPH 3; Rome, 1898).

Acta Capitulorum Provincialium Provinciae Lombardiae, ed. in Thomas Kaeppeli, 'Acta Capitulorum Provinciae Lombardiae (1254–1293) et Lombardiae Inferioris (1309–1312)', *AFP* 11 (1941), 140–67.

Acta Capitulorum Provincialium Provinciae Romanae (1243–1344), ed. Thomas Kaeppeli and Antoine Dondaine (MOFPH 20; Rome, 1941).

ALBERIC OF TROIS-FONTAINES, *Chronica*, ed. P. Scheffer-Boichorst (MGH.Ss. 23; Hanover, 1904).

ALBERTI, LEANDRO, *De Viris Illustribus Ordinis Praedicatorum* (Ferrara, 1516).

ALESSANDRINO, GILSBERTO, RECUPERATO OF PIETRAMALA; PAPPARONI, ALDOBRANDINO, and VISDOMNI, OLDRADO, *Legenda Antica Beati Ambrosii Senensis*, ed. Daniel Papabroch (AS 8: Mar. III; Paris, 1865), 180–209.

ANDRÉ OF FONTEVRAULT, *Alia Vita Beati Roberti [de Arbrissello]*, ed. Joannes Bollandus. (AS 6: Feb. III; Paris, 1865), 613–21.

Annales Bergomates 1156–1266, ed. Oswald Holder-Egger (MGH.Ss. 31; Hanover, 1904), 323–35.

Annales Brixienses, ed. Ludwig Bethmann (MGH.Ss. 18; Hanover, 1863), 811–20.

Annales Parmenses Maiores, ed. Georg Heinrich Pertz (MGH.Ss. 18; Hanover, 1863), 790–9.

Annales Placentini Guelfi, ed. Georg Heinrich Pertz (MGH.Ss. 18; Hanover, 1863), 411–57.

Annales Veteres Mutinenses ab Anno MCXXXI usque ad MCCCXXXVI (RIS 11, cols. 53–130).

Annales Veteres Veronenses, ed. Carlo Cipolla, *Archivio Veneto*, 9 (1875), 77–98.

ANONYMUS PERUSINUS, *Legenda*, in *Scripta Leonis, Rufini et Angeli Sociorum S. Francisi: The Writings of Leo, Rufino and Angelo, Companions of St. Francis*, ed. Rosalind B. Brooke (Oxford, 1970), 86–291.

ANTHONY OF SIENA, *Chronica Fratrum Ordinis Praedicatorum* (Paris, 1585).

Antiquitates Italicae Medii Aevi, iv, ed. Lodovico Antonio Muratori (Milan, 1741).

ANTONINUS OF FLORENCE, *Chronicorum Opus* (3 vols.; Lyon, 1586).

Assidua, see *Vita prima di s. Antonio*.

AUVRAY, LUCIEN, see GREGORY IX.

BARTHOLOMEW OF PISA, *De Conformitate Vitae Beati Francesci ad Vitam Domini Iesu* (AF 4–5, 2 vols.; Quaracchi, 1912–17).

BERNARD OF BESSA, *Liber de Laudibus Beati Francisci*, AF 4. Quaracchi, 1912).

BERNARD OF CLAIRVAUX, *Sermones super Cantica Canticorum*, 2 vols.; ed. J. Leclercq *et al.* (Sancti Bernardi Opera, 1–2; Rome, 1957–8).

BOLOGNETTI, *Cronaca detta dei Bolognetti*, in CCB 2, ed. Albano Sorbelli (RIS² 18:1:2, Città di Castello, 1911).

BONATTI, GUIDO, *De Astronomia Tractatus X*, ed. Nicolaus Prukner (Basel, 1550).

BONCAMPAGNO OF SIGNI, *Testi riguardanti la vita degli studenti a Bologna nel sec. XIII (dal* Boncompagnus, *lib. I)*, ed. Vigilio Pini (Testi per esercitazioni accademiche, 6; Bologna, 1968).

BORSELLI, GIROLAMO ALBERTUCCI DE', *Cronica Gestorum ac Factorum Memorabilium Civitatis Bononie*, ed. Albano Sorbelli (RIS² 23:2; Città di Castello, 1912).

Bullarium Franciscanum Romanorum Pontificum, i, ed. Giovànni Giacinto Sbaraglia (7 vols.; Rome, 1759).

Bullarium Ordinis Fratrum Praedicatorum, i. ed. Thomas Ripoll (Rome, 1729).

CANTINELLI, PIETRO, *Chronicon [AA 1228–1306]*, ed. Francesco Torraca (RIS² 28; Città di Castello, 1902).

Chronica XXIV Generalium Ordinis Minorum (AF 3; Quaracchi, 1894).

Chronicon Affligemense, ed. Georg Waitz (MGH.Ss. 9; Hanover, 1851), 407–17.

Chronicon Estense cum Additamentis usque ad Annum 1479, ed. Giulio Bertani and Emilio Paolo Vicini (RIS² 15; Città di Castello, 1908).

Chronicon Faventinum Magistri Tolosani [AA 20 av. c.-1236], ed. Giuseppe Rossini (RIS² 28; Bologna, 1939).

Chronicon Marchiae Trevisinae et Lombardiae, ed. L. A. Botteghi (RIS² 8:3; Città di Castello, 1916).

Chronicon Parmense ab Anno 1038 usque ad Annum 1338, ed. Giuliano Bonazzi (RIS² 9:9; Città di Castello, 1902).

Chronicon Placentinum, ed. Lodovico Antonio Muratori (RIS xvi).

CONSTANTINE OF ORVIETO, *Legenda Sančti Dominici*, ed. H. C. Sheeben (MOFPH 16: MHD 2; Rome, 1935), 263–353.

Dominican Constitutions 1228. 'Die Constitutionen der Predigerordens vom Jahre 1228', ed. H. Denifle, *Archiv für Literatur- und Kirchengeschichte des Mittelalters*, i (1885), 165–227.

Die Doppelchronik von Reggio und die Quellen Salimbenes, ed. Alfred Wilhelm Dove (Leipzig, 1873).

Dossier de l'ordre de la pénitence au XIIIe siècle, ed. Gilles-Gérard Meersseman (Spicilegium Friburgense, 7; Fribourg, 1961).

Epistolae Saeculi XIII ex Registris Pontificum Romanorum, ed. Carl Rodenberg (3 vols.; Berlin, 1883–94).

FIAMMA, GALVANO, *Cronica Maior Ordinis Praedicatorum*, ed. in Gundisalvo Odetto, 'La Cronaca maggiore dell'Ordine Domenicano di Galvano Fiamma: Frammenti inediti', *AFP* 10 (1940), 319–73.

—— *Chronica [Parva] Ordinis Praedicatorum*, ed. Benedictus Maria Reichert (MOFPH 2; Rome, 1896).

—— *Manipulus Florum*, ed. Lodovico Antonio Muratori (RIS 11; Milan, 1729), 537–739.

I Fioretti di san Francesco, ed. Paul Sabatier (Fonti Francescane, 1; Assisi, 1970).

St Francis of Assisi, Writings and Early Biographies: English Omnibus of the Sources for the Life of St. Francis, ed. Marion A. Habig (Chicago, 1972).

GEOFFROY LE GROS, *Vita Beati Bernardi [Tironiensis]*, ed. Godefridus Henschenius (AS 11: Apr. II; Paris, 1866), 222–54.

GEOFFROY OF VENDÔME, 'Epistola Fratri Roberto' (Epistola 47), *Goffridi Abbatis Epistolae*, PL 157:181–4.

GODI, ANTONIO, *Cronaca*, ed. Giovanni Soranzo (RIS² 8:1; Città di Castello, 1905).

GREGORY THE GREAT, *Dialogues*, ed. Adalbert de Vogüe (Sources Chrétiennes; Paris, 1979).

GREGORY IX, *Les Registres de Grégoire IX: Recueil des bulles de ce pape*, i, ed. Lucien Auvray (Paris, 1896).

GRIFFONI, MATTEO, *Memoriale Historicum de Rebus Bononiensium*, ed. Lodovico Frati and Albano Sorbelli (RIS² 8:2; Città di Castello, 1902).

GUIBERT OF NOGENT, *Gesta Dei per Francos*, PL 156.

HUGH OF SAINT-CHER, *Théorie de la prophétie et philosophie de la connaissance aux environs de 1230: La Contribution d'Hugues de Saint-Cher (MS Douai 434, Question 481)*, ed. Jean-Pierre Torrell (Louvain, 1977).

HUILLARD-BRÉHOLLES, JEAN LOUIS ALPHONSE (ed.), *Historia Diplomatica Friderici Secundi* (7 vols.; Paris, 1852–61).

JACQUES DE VITRY, *The* Historia Occidentalis *of Jacques de Vitry: A Critical Edition*, ed. J. F. Hinnebusch (Spicilegium Friburgense, 17; Fribourg, 1972).

JOACHIM OF FIORE, *Expositio in Apocalypsim* (Venice, 1527).

—— *Il libro delle figure dell'Abate Gioachino da Fiore*, ed. Luigi Tondelli (2 vols.; Turin, 1953).

JORDAN OF SAXONY, *Liber de Principiis Ordinis Praedicatorum*, ed. H.-C. Sheeben (MOFPH 16: MHD 2; Rome, 1935), 1–88. (Conventionally known as the *Libellus*).

Legend of Perugia, see Anonymus Perusinus.

Liber Regiminum Padue, ed. Antonio Bonardi (RIS² 8:1; Città di Castello, 1905), 269–376.

MARBOD OF RENNES, *Marbodi Redonensis Episcopi Epistolae*, (Epistola 6), PL 171:1480–8.

MATTHEW PARIS, *Chronica Maiora*, ed. Henry Richard Luard (Rolls Series, 57; 7 vols.; London, 1872–83).

MAURISIO, GERARDO, *Cronica Dominorum Ecelini et Alberici Fratrum de Romano*, ed. Giovanni Soranzo (RIS² 8:4; Città di Castello, 1914).

Memoriae Mediolanenses, ed. Philipp Jaffé (MGH.Ss. 18; Hanover, 1863), 399–402.

Memoriale Potestatum Regiensium, ed. Lodovico Antonio Muratori (RIS 8, cols. 1071–80).

MILIOLI, ALBERTO, *Liber de Temporibus et Aetatibus et Cronica Imperatorum*, ed. Oswald Holder-Egger (MGH.Ss. 31; Hanover, 1904), 336–668.

MUSSI, GIOVANNI DE', *Chronicon Placentinum*, ed. Lodovico Antonio Muratori (RIS 16).

MUZIO OF MODENA, *Annales Placentini Gibellini*, ed. Georg Heinrich Pertz (MGH.Ss. 18; Hanover, 1863), 457–581.

PARISIO OF CEREA, *Annales Veronenses*, ed. Philipp Jaffé (MGH.Ss. 19; Hanover, 1866), 2–18.

PIETRO GERARDO, *Vita et gesti di Ezzelino terzo da Romano* (Miscellanea di storia veneta, 2nd. ser., 2; Venice(?), 1894).

Poésies populaires latines du moyen âge, ed. Édélstand du Méric (Paris, 1847).

Rampona. *Cronaca A detta volgarmente Rampona*, in *CCB* 2, ed. Albano Sorbelli (RIS^2 18:1:2; Città di Castello, 1911).

Regesten des Kaiserreichs unter Philipp, Otto IV., Friedrich II., Heinrich (VII.), Conrad IV., Heinrich Raspe, Wilhelm und Richard, 1198–1272, ed. Julius Ficker and Eduard Winkelmann. Regista Imperii, 5. (Innsbruck, 1901).

REMIGIO DE' GEROLAMI, *De Bono Pacis*, ed. in Maria Consiglia De Matteis, *La 'teologia politica comunale' di Remigio de' Girolami*, (Il mondo medievale, 3; Bologna, 1977), 53–72.

RICHARD OF SAN GERMANO, *Chronica*, ed. Carlo Alberto Garufi (RIS^2 7:2; Bologna, 1935–7).

ROLANDINO OF PADUA, *Cronica in Factis et circa Facta Marchie Trivixiane*, ed. Antonio Bonardi (RIS^2 8:1; Città di Castello, 1905).

SALIMBENE DE ADAM, *Cronica*, ed. Oswald Holder-Egger (MGH.Ss. 32; Hanover, 1905–13); trans. Joseph L. Baird, Giuseppe Baglivi, and John Robert Kame, *The Chronicle of Salimbene de Adam* (Medieval and Renaissance Texts and Studies, 40; Binghamton, NY, 1986).

SAVIOLI, LODOVICO VITTORIO, *Annali bolognesi* (6 vols.; Bessano, 1784–95).

SIGONIO, CARLO, *Historia de Rebus Bononiensibus Libri VIII* (Frankfurt, 1604).

Speculum Perfectionis, in *Le Speculum Perfectionis; ou, Mémoires de frère Léon sur la seconde partie de la vie de saint François d'Assise*, ed. Paul Sabatier (2 vols.; Manchester, 1922–31).

TAEGIO, AMBROGIO, *Legenda Beatissimi Petri Martyris ex Multis Legendis in Unum Compilata* (AS, 12: Apr. III; Paris, 1867), 694–727.

THOMAS OF CANTIMPRÉ, *Bonum Universale de Apibus*, ed. in part Joannes Baptista Sollerius (AS, 28: Jul. I; Paris, 1867), 424–5.

THOMAS OF PAVIA, *Dialogus de Gestis Sanctorum Fratrum Minorum*, ed. Ferdinand M. Delorme, (Bibliotheca Franciscana Ascetica Medii Aevi, 5; Quaracchi, 1923).

THOMAS OF SPALATO, *Historia Pontificum Salonitanorum et Spalatinorum*, ed. L. De Heinemann (MGH.Ss. 24; Hanover, 1892), 568–98.

Tonini, Luigi, *Rimini nel secolo xiii*, vol. iii of *Della storia civile e sacra riminese* (Rimini, 1862).

Varignana. *Cronaca B detta volgarmente Varignana*, in *CCB* 2, ed. Albano Sorbelli (RIS² 18:1:2; Città di Castello, 1911).

Vignati, Cesare, *Storia diplomatica della Lega Lombarda* (Milan, 1867, repr. Turin, 1966).

Villola, Pietro and Floriano da, *Cronaca*, in *CCB* 2, ed. Albano Sorbelli (RIS² 18:1:2; Città di Castello, 1911).

Visconti, Federico, 'Sermones Frederici de Vicecomitibus, Archiepiscopi Pisani de S. Francesco', ed. M. Bihl, *Archivum Franciscanum Historicum*, 1 (1908), 652–5.

Vita prima di s. Antonio o 'Assidua' (c.1232), ed. and trans. Vergilio Gamboso (Fonti agiografiche antoniane, 1; Padua, 1981).

Vita Sancti Norberti Archiepiscopi Magdeburgensis {B}, ed. Daniel Papebroch (AS 21: Jun. I; Paris, 1867), 807–45.

Vitae Fratrum Ordinis Praedicatorum, ed. Benedictus Maria Reichert (MOFPH 1; Louvain, 1896).

Wadding, Luke, *Annales Minorum seu Trium Ordinum S. Francisco Institutorum* (25 vols.; Rome, 1731–1886).

William of Saint-Amour, *De Periculis Novissimi Temporis*, ed. Max Bierbaum as *Bettelorden und Weltgeistlichkeit an der Universität Paris* (Münster, 1920).

William of Saint-Thierry, *Sancti Bernardi Abbatis Clarae-Vallensis Vita et Res Gestae*, PL 185:222–416.

SECONDARY WORKS

Abate, Giuseppe, 'Il "Liber Epilogorum" di fra' Bartolomeo da Trento O.P. in due codici rintracciati nella Biblioteca Antoniana di Padova', in *Miscellanea Pio Paschini* (Rome, 1948), i. 269–92.

—— and Luisetto, Giovanni, *Codici e manoscritti della Biblioteca Antoniana* (3 vols.; Vicenza, 1975).

Abulafia, David, *Frederick II: A Medieval Emperor* (London, 1988).

Affò, Ireneo, *Storia di Parma* (4 vols.; Parma, 1793).

Archivio Sartori (Padua, 1984).

Arnaldi, Girolamo, *Studi sui cronisti della Marca trevigiana nell'età di Ezzelino da Romano* (Rome, 1963).

Aron, Marguerite, *St Dominic's Successor* (London, 1955).

Bader, Karl S., 'Arbiter, Arbitrator seu Amicabilis Compositor', *Zeitschrift der Savigny-Stiftung für Rechtsgeschichte (Kan. Abt.)*, 77 (1960), 239–76.

Barbarano, Francesco, *Historia ecclesiastica della città, territorio e diocesi di Vicenza* (Vicenza, 1649).

BARONE, GIULIA, 'L'Ordine dei Predicatori e le città: Teologia e politica nel pensiero e nell'azione dei predicatori', *MEFR* 89 (1977), 609–17.

BATAILLON, LOUIS-JACQUES, 'La predicazione dei religiosi mendicanti del secolo XIII nell'Italia centrale', *MEFR* 89 (1977), 691–4.

[BECCARI, TOMMASO], *Lettere di Giovanni Domenico dei Coppa [pseud.] in difesa di fra Giovanni da Vicenza domenicano, celebre predicatore del secolo XIII* (Nuova raccolta d'opuscoli scientifici e filologici, 16–17, 2 vols.; Venice, 1785–7).

BEDOUELLE, GUY, *Dominique ou la grâce de la parole* (Belgium, 1982).

BÉRIOU, NICOLE, 'L'Art de convaincre dans la prédication de Randolphe d'Homblières', in *Faire croire: Modalités de la diffusion et de la réception des messages religieux du XIIᵉ au XVᵉ siècle* (Collection de l'École Française de Rome, 51; Rome, 1981), 39–65.

BESTA, ENRICO, *Fonti, legislazione e scienza giuridica dalla caduta dell'Impero Romano al secolo decimosesto*, in Pasquale del Giudice (ed.), *Storia di diritto italiano*, i(2), (Florence, 1969).

BOLTON, BRENDA, 'Innocent III's Treatment of the Humiliati', in C. J. Cuming and Derek Baker (ed.), *Popular Belief and Practice* (Studies in Church History, 8; Cambridge, 1972), 73–82.

BONNIWELL, WILLIAM R., *The History of the Dominican Liturgy* (New York, 1944).

BORTOLAMI, SANTE, 'La città del santo e del tiranno: Padova nel primo duecento', in *S. Antonio 1231–1981: Il suo tempo, il suo colto, e la sua città* (Padua, 1981), 244–61.

BORTOLAN, DOMENICO, *Santa Corona: Chiesa e convento dei domenicani in Vicenza, memorie storiche* (Vicenza, 1889).

BOWSKY, WILLIAM, 'The Medieval Commune and Internal Violence: Police Power and Public Safety in Siena, 1287–1355', *American Historical Review*, 73 (1967), 1–17.

——— *A Medieval Italian Commune: Siena Under the Nine, 1287–1355* (Berkeley, Calif., 1981).

BRENTANO, ROBERT, *Two Churches: England and Italy in the Thirteenth Century*, 2nd edn. (Berkeley, Calif., 1988).

BREZZI, PAOLO, *I comuni cittadini italiani: Origine e primitiva costituzione* (Milan, 1940).

——— 'I comuni cittadini italiani e l'Impero medioevale', in *Nuove questioni di storia medioevale* (Milan, 1969), 177–208.

BROWN, DANIEL A., 'The Alleluia: A Thirteenth Century Peace Movement', *Archivum Franciscanum Historicum*, 81 (1988), 3–16.

BROWN, PETER, 'The Rise and Function of the Holy Man in Late Antiquity', *Journal of Roman Studies*, 61 (1971), 80–101.

——— 'Town, Village and Holy Man: The Case of Syria', in *Assimilation et résistance à la culture gréco-romaine dans le monde ancien* (Paris, 1976), 213–20.

——— 'Society and the Supernatural: A Medieval Change', in *Society and the Holy in Late Antiquity* (Berkeley, Calif., 1982), 302–32.

BRUCKER, GENE, *Florentine Politics and Society, 1343–1378* (Princeton, NJ, 1962).

BUCKLAND, W. W., *A Textbook of Roman Law from Augustus to Justinian*, 2nd edn. (Cambridge, 1950).

BUTLER, W. F., *The Lombard Communes* (New York, 1906).

CALVI, PAOLO, 'Angelogabriello di Santa Maria', in *Biblioteca e storia di scrittori di Vicenza*, i (Vicenza, 1772).

CANTINI, GUSTAVO, 'L'apostolato dei beati Gherardo Boccabadati e Leone Valvassori da Perego francescani e la devozione dell'Alleluia', *SF*, 3rd ser., 9 (1938), 335–53.

CAPO, LIDIA, 'Cronache mendicanti e cronache cittadine', *MEFR* 89 (1977), 633–9.

CASTAGNETTI, ANDREA, 'Appunti per una storia sociale e politica delle città della Marca veronese-trevigiana (secoli XI–XIV)', in Reinhard Elze and Gina Fasoli (ed.), *Aristocrazia cittadina e ceti popolari nel tardo medioevo in Italia e Germania* (Annali dell'Istituto Storico Italo-Germanico, 13; Bologna, 1984), 41–78.

CHARLAND, THOMAS MARIE, '*Praedicator Gratiosus*', *Revue dominicaine* (Ottawa), 39 (1933), 88–96.

CIPOLLA, CARLO, *La storia politica di Verona* (Verona, 1954).

CIPOLLA, CARLO, 'L'omiletica nel medioevo: Teoria sociale e comunicazione di massa', *Verifiche*, 6 (1977), 298–360.

COLETTI, VITTORIO, *Parole dal pulpito: Chiesa e movimenti religiosi tra latino e volgare nell'Italia del medioevo e del rinascimento* (Casale Monferrato, 1983).

CORIO, BERNARDINO, *Mediolanensis Patria Historia* (Milan, 1503).

—— *Storia de Milano* (3 vols.; Milan, 1855–7).

COULTON, G. G., *From St Francis to Dante: Translations from the Chronicle of the Franciscan Salimbene (1221–1288) with Notes and Illustrations from other Medieval Sources*, 2nd. edn. (1907; repr. Philadelphia, 1972).

CRACCO, GIORGIO, 'Da comune di famiglia a città satellite (1183–1311)', in *Storia di Vicenza*, ii: *L'età medievale*, ed. Giorgio Cracco (Vicenza, 1988), 73–283.

CREYTENS, RAYMOND, 'Les Écrivains dominicains dans la chronique d'Albert de Castello (1516)', *AFP* 30 (1960), 227–313.

CUSIN, F., 'Per la storia del castello medievale', *Revista storica italiana*, 7th. ser., 20 (1887), 25–58, 178–204.

D'ALATRI, MARIANO, 'Predicazione e predicatori francescani nella *Cronica* di fra' Salimbene', *Collectanea francescana*, 46 (1976), 63–91.

D'AMATO, ALFONSO, *La devozione a Maria nell'Ordine Domenicano* (Bologna, 1984).

D'AVRAY, D. L. *The Preaching of the Friars: Sermons Diffused from Paris before 1300* (Oxford, 1985).

DELCORNO, CARLO, *Giordano da Pisa e l'antica predicazione volgare* (Florence, 1974).

—— *La predicazione nell'età comunale* (Sansoni Scuola Aperta, 57; Florence, 1974).

—— 'Origini della predicazione francescana', in *Francesco d'Assisi e francescanesimo dal 1216 al 1226* (Atti del IV Convegno Internazionale, Assisi, 15–17 ottobre 1976; Assisi, 1977), 125–60.

—— 'Predicazione volgare e volgarizzamenti', *MEFR* 89 (1977), 679–89.

—— 'Rassegna di studi sulla predicazione medievale e umanistica', *Lettere italiane*, 33 (1981), 235–76.

DENIFLE, HEINRICH, *Die Universitäten des Mittelalters bis 1400* (Berlin, 1885).

DICKERHOF, HARALD, 'Friede als Herrschaftslegitimation in der italienischen Politik des 13. Jahrhunderts', *Archiv für Kulturgeschichte*, 59 (1977), 366–89.

DICKSON, GARY, 'The Flagellants of 1260 and the Crusades', *Journal of Medieval History*, 15 (1989), 227–67.

DONDAINE, ANTOINE, 'Saint-Pierre-Martyr: Études', *AFP* 23 (1953), 66–162.

DU CANGE, CHARLES, *Glossarium Mediae et Infimae Latinitatis* (11 vols.; Paris, 1937–8).

Early Dominicans: Selected Writings, ed. Simon Tugwell (The Classics of Western Spirituality; New York, 1982).

ELIADE, MIRCEA, *Shamanism: Archaic Techniques of Ecstasy* (London, 1964).

L'eremitismo in occidente nei secoli XI e XII (La Seconda Settimana Internazionale di Studio, Mendola, 30 August–6 September, 1962; Milan, 1965).

FASOLI, GINA, 'Le compagnie delle armi a Bologna', *L'Archiginnasio*, 28 (1933), 158–83; 323–40.

—— 'Le compagnie delle arti a Bologna fino al principio del secolo XV', *L'Archiginnasio*, 30 (1935), 237–80.

—— (ed.), *Studi ezzeliniani* (Rome, 1963).

—— 'Gouvernants et gouvernés dans les communes italiennes du XIe au XIIIe siècle', *Receuils de la Société Jean Bodin*, 25 (1965), 47–96.

—— 'Oligarchia e ceti popolari nelle città padane fra il XIII e il XIV secolo', in Reinhard Elze and Gina Fasoli (ed.), *Aristocrazia cittadina e ceti popolari nel tardo medioevo in Italia e Germania* (Annali dell'Istituto Storico Italo-Germanico, 13; Bologna, 1984), 11–40.

—— 'Città e feudalità', in *Structures féodales et féodalisme dans l'occident méditerranéen IXe–XIIIe siècles* (Rome, 1986), 380–2.

FONTANA, VINCENZO MARIA, *Monumenta Dominicana* (Rome, 1675).

FORNI, A., 'Kerygma e adattamento: Aspetti della predicazione cattolica nei secoli XII e XIV', *BISI* 89 (1980/81), 261–348.

—— 'La "Nouvelle Prédication" des disciples de Foulques de Neuilly: Intentions, techniques et réactions', in *Faire croire: Modalités de la diffusion et de la réception des messages religieux du XIIe au XVe siècle* (Collection de l'École Française de Rome, 51; Rome, 1981), 19–37.

FORTE, STEFANO L., 'Le province domenicane in Italia nel 1650', *AFP* 41 (1971), 325–458.

FOWLER, LINDA, 'Forms of Arbitration', in Stephen Kuttner (ed.), *Proceedings of the Fourth International Congress of Medieval Canon Law (Toronto, 21–25 August 1972)* (Vatican City, 1976), 133–47.

FRANCHINI, VITTORIO, *Saggio di ricerche sull'istituto del podestà nei comuni medievali* (Bologna, 1912).

FUMAGALLI, VITO, 'In margine all' "Alleluia" del 1233', *BISI* 80 (1968), 257–72.

GARBETT, G. KINGSLEY, 'Spirit Mediums as Mediators in Korekore', in John Beattie *et al.* (ed.) *Spirit Mediumship and Society in Africa* (New York, 1969), 104–27.

GATTI, DANIELA, 'Religiosità populare e movimento di pace nel'Emilia del secolo XIII', in *Itinerari storici: Il medioevo in Emilia* (Carpi, 1980), 79–107.

GATTO, LODOVICO, 'Il sentimento cittadino nella *Chronica* di Salimbene', in *La coscienza cittadina nei comuni italiani del duecento, 11–14 ottobre 1970* (Convegni del Centro di Studi sulla Spiritualità Medievale, 11; Todi, 1972), 365–94.

GAUDENZI, AUGUSTO, 'Gli statuti delle Società delle Armi del Popolo di Bologna', *Bollettino dell'Istituto Storico Italiano*, 8 (1889), 7–74.

GENNEP, A. VAN, *The Rites of Passage*, trans. M. B. Vizedom and G. L. Caffee (Chicago, 1966).

GOODRICH, MICHAEL, 'The Politics of Canonization in the Thirteenth Century: Lay and Mendicant Saints', *Church History*, 44 (1975), 294–307.

HANAUER, G. 'Das Berufspodestat im 13. Jahrhundert', *Mittheilungen des Instituts für österreichishe Geschichtsforschung*, 23 (1902), 378–426.

HEFELE, HERMANN, *Die Bettelorden und das religiöse Volksleben Ober- und Mittelitaliens im XIII. Jahrhundert* (Leipzig, 1910).

HELMHOLZ, R. H., *Marriage Litigation in Medieval England* (Cambridge, 1974).

HESSEL, ALFRED, *Storia della città di Bologna dal 1116 al 1280*, ed. and trans. Gina Fasoli (Bologna, 1975).

HOUSLEY, N. J., 'Politics and Heresy in Italy: Anti-Heretical Crusades, Orders and Confraternities', *Journal of Ecclesiastical History*, 33 (1982), 193–208.

HYDE, J. K., 'Contemporary Views on Faction and Civil Strife in Thirteenth and Fourteenth Century Italy', in Lauro Martines (ed.), *Violence and Civil Disorder in Italian Cities, 1200–1500*. (Berkeley, Calif., 1972), 273–307.

—— *Society and Politics in Medieval Italy: The Evolution of the Civic Life, 1000–1350* (New Studies in Medieval History; London, 1973).

KAEPPELI, THOMAS, 'Der literarische Nachlaß des sel. Bartholomaeus von Vicenza O.P.', in *Mélanges Auguste Pelzer* (Université de Louvain, Recueil de Travaux d'Histoire et de Philologie, 3rd ser., 26; Louvain, 1947), 275–301.

—— 'Le prediche del b. Ambrogio Sansedoni da Siena', *AFP* 38 (1968), 5–12.

KANTOROWICZ, ERNST, *Frederick the Second, 1194–1250*, trans. E. O. Lorimer (New York, 1957).

KEE, HOWARD CLARK, *Miracle in the Early Christian World: A Study in the Socio-Historical Method* (New Haven, 1983).

KOUDELKA, VLADIMIR J., 'La fondazione del convento domenicano di Como (1233–1240)', *AFP* 36 (1966), 395–427.

—— 'Procès de canonisation de s. Dominique', *AFP* 42 (1972), 47–67.

KUCZYNSKI, JOSEPH, *Le Bienheureux Guala de Bergame de l'Ordre des Frères Prêcheurs* (Estavayer, 1916).

KUEHN, THOMAS, 'Arbitration and Law in Renaissance Florence', *Renaissance and Reformation* 23 (1987), 289–319.

LARNER, JOHN, *The Lords of the Romagna: Romagnol Society and the Origins of the Signorie* (Ithaca, NY, 1965).

LEA, HENRY CHARLES, *History of the Inquisition in the Middle Ages* (3 vols.; London, 1886).

LEICHT, PIER S., 'La corporazione italiana delle arti nelle sue origini e nel primo periodo del comune', *Scritti vari di storia del diritto* (Milan, 1943), i. 297–308, 431–48.

LESNICK, DANIEL R., *Preaching in Medieval Florence: The Social World of Franciscan and Dominican Spirituality* (Athens, Ga., 1989).

LITTLE, LESTER K., 'Evangelical Poverty, the New Money Economy and Violence', in David Flood (ed.), *Poverty in the Middle Ages* (Werl, 1975), 11–29.

LIUZZI, F., *La lauda e i primordi della melodia italiana* (2 vols.; Rome, 1934).

LONGÈRE, JEAN, 'Le Pouvoir du prêcheur et le contenu de la prédication dans l'occident chrétien', in *Prédication et propagande au moyen âge: Islam, Byzance, Occident* (Penn–Paris–Dumbarton Oaks Colloquia, 3; Paris, 1980), 165–78.

—— *La Prédication médiévale* (Paris, 1983).

McLOUGHLIN, WILLIAM G., jr., *Modern Revivalism: Charles Grandison Finney to Billy Graham* (New York, 1959).

—— *Revivals, Awakenings, and Reform: An Essay on Religion and Social Change in America, 1607–1977* (Chicago, 1978).

MAGLI, IDA, 'Un linguaggio di massa del medioevo: L'oratoria sacra', *Rivista di sociologia*, 1 (1963), 181–98.

—— *Gli uomini della penitenza: Lineamenti antropologici del medioevo italiano* (Milan(?), 1967).

MAGRINI, ANTONIO, *Notizie di fra Giovanni da Schio per le nozze di Nanne Gozzadini e Maria Teresa Sarego* (Padua, 1841).

MAIOCCHI, RODOLFO, *Il beato Isnardo da Vicenza O.P. e il suo apostolato in Pavia nel secolo XIII* (Pavia, 1910).

MANSELLI, RAOUL, 'Ezzelino da Romano nella politica italiana del secolo XIII', *Studi ezzeliniani* (Rome, 1963), 35–75.

—— *La Religion populaire au moyen âge: Problèmes de méthode et d'histoire* (Montreal, 1975).

MANTEUFFEL, TADEUSZ, *Naissance d'une hérésie: Les Adeptes de la pauvreté volontaire au moyen âge* (Civilisation et sociétés, 6; Paris, 1970).

MARETO, FELICE DA, 'L'ordine francescano della penitenza a Parma, Fidenza, Piacenza e Modena', in Mariano d'Alatri (ed.), *Il movimento francescano della penitenza nella società medioevale: Atti del Terzo Convegno di Studi Francescani, Padova, 25—26—27 settembre 1979* (Rome), 311–21.

MARTINES, LAURO, 'Political Conflict in the Italian City States', *Government and Opposition*, 3 (1968), 69–91.

—— 'Political Violence in the Thirteenth Century', in id. (ed.), *Violence and Civil Disorder in Italian Cities, 1200–1500* (Berkeley, Calif., 1972), 331–5.

MARTONE, LUCIANO, Arbiter-Arbitrator: *Forme di giustizia privata nell'éta del diritto comune* (Naples, 1984).

MEERSSEMAN, GILLES-GÉRARD, 'Les Confréries de Saint-Pierre-Martyr', *AFP* 21 (1951), 51–196.

—— 'Les Milices de Jésus-Christ', *AFP* 23 (1953), 275–308.

—— 'Disciplinati e penitenti nel duecento', in *Il movimento dei Disciplinati nel settimo centenario dal suo inizio (Perugia-1260)* (Convegno Internazionale, Perugia, 25–28 settembre 1960; Perugia, 1962), 43–72.

—— Ordo Fraternitatis: *Confraternite e pietà dei laici nel medioevo* (3 vols.; Italia sacra, 24–6; Rome, 1977).

MERLO, GRADO G., *Eretici e inquisitori nella società piemontese del trecento* (Turin, 1977).

MONTI, GENNARO M., *Le confraternite dell'alta e media Italia* (2 vols.; Venice, 1927).

MORGHEN, RAFFAELLO, 'Ranieri Fasani e il movimento dei Disciplinati del 1260', in *Il movimento dei Disciplinati nel settimo centenario dal suo inizio (Perugia-1260)* (Convegno Internazionale, Perugia, 25–28 settembre 1960; Perugia, 1962), 29–42.

MORRIS, COLIN, *The Discovery of the Individual, 1050–1200* (London, 1972).

MOSCHETA, VALERIO, B. *Joannis Vicentii O.P., Professi Cenobii Paduani S. Augustini, Doctrina, Sanctitate, et Miraculis Insignis Praeclara Gesta* (Padua, 1590), ed. Joannes Baptista Sollerius (AS 28: Jul. I; Paris, 1867), 412–24.

Il movimento dei Disciplinati nel settimo centenario dal suo inizio (Perugia, 1260) (Convegno Internazionale, Perugia, 25–28 settembre 1960; Perugia, 1962).

MURRAY, ALEXANDER, 'Piety and Impiety in Thirteenth Century Italy', in C. J. Cuming and Derek Baker (ed.), *Popular Belief and Practice* (Studies in Church History, 8; Cambridge, 1972), 83–106.

NICCOLAI, FRANCO, *Contributo allo studio dei più antichi brevi della compagna genovese* (Milan, 1939).

—— 'I consorzi nobiliari ed il comune nell'alta e media Italia', *Revista di storia del diritto italiano*, 13 (1940), 116–47, 292–341, 397–477.

NICOLA, P. DI, 'Omelia come strumento di comunicazione di massa', *Sociologia*, NS, 10 (1976), 1979–97.

ODETTO, GUNDISALVO, 'La Cronaca maggiore dell'Ordine Domenicano di Galvano Fiamma: Frammenti inediti', *AFP* 10 (1940), 297–373.

ORLANDINI, GIANFRANCO, '"Studio" e scuola di notariato', *Atti del Convegno Internazionale di Studi Accursiani: Bologna, 21–26 ottobre 1963* (Milan, 1968), 73–95.

PAGLIARINI, GIAMBATTISTA, *Cronaca di Vicenza*, ed. and trans. (from unpublished Latin original listed with manuscripts) G. Alcaini (Vicenza, 1663).

PALTRINIERI, I., 'Un nuovo codice di fra' Bartolomeo tridentino', *Aevum*, 20 (1946), 3–13.

—— and SANGALLI, G., 'Un opera finora sconosciuta: *Il Liber Miraculorum B.V.M.* di fra' Bartolomeo tridentino', *Salesianum*, 12 (1950), 372–97.

PASQUALI, GIANFRANCO, 'Gerolamo Albertucci de' Borselli, O.P. (1432–1497): Ricerche bio-bibliografiche', *Rivista di storia della Chiesa in Italia*, 25 (1971), 59–82.

PATLAGEAN, EVELYNE, 'Sainteté et pouvoir', in Sergei Hackel (ed.), *The Byzantine Saint*, (University of Birmingham Fourteenth Spring Symposium of Byzantine Studies; London, 1981), 88–105.

PERELLA, NICOLAS JAMES, *The Kiss Sacred and Profane: An Interpretative History of Kiss Symbolism and Related Religio-Erotic Themes* (Berkeley, Calif., 1969).

PETERS, EDWARD, *The Magician, the Witch and the Law* (Philadelphia, 1978).

PETRICCIONE, VINCENZO, 'I penitenti francescani di Bologna nel secolo XIII', in Mariano d'Alatri (ed.), *I frati penitenti di san Francesco nella società del due- e trecento* (Secondo Convegno di Studi Francescani, Roma, 12–13–14 ottobre; Rome, 1977), 259–69.

PHARR, MARY BROWN, 'The Kiss in Roman Law', *Classical Journal* 42 (1946/47), 393–7.

PILONI, GEORGIO, *Historia, nelle quale, oltre molte cose degne, avvenute in diverse parti del mondo di tempo in tempo, s'intendono e leggono d'anno in anno con minuto reguaglio tutti i successi nella città di Belluno* (Venice, 1607).

PINI, ANTONIO IVAN, *Città, comuni e corporazioni nel medioevo italiano* (Bologna, 1986).

PIÒ, GIOVANNI MICHELE, *Vite degli huomini illustri di s. Domenico* (Bologna, 1620).

PIVANO, SILVIO, *Stato e chiesa negli statuti comunali italiani* (Turin, 1904).

POGGIALI, CRISTOFORO, *Memorie storiche di Piacenza* (12 vols.; Piacenza, 1757–66).

QUÉTIF, JACQUES and ÉCHARD, JACQUES, *Scriptores Ordinis Praedicatorum* (2 vols.; Paris, 1719).

REEVES, MARJORIE, *The Influence of Prophecy in the Later Middle Ages: A Study in Joachimism* (Oxford, 1969).

—— *Joachim of Fiore and the Prophetic Future* (London, 1976).

—— and HIRSCH-REICH, BEATRICE, *The Figurae of Joachim of Fiore* (Oxford, 1972).

RENARD, J. P., *La Formation et la désignation des prédicateurs au début de l'Ordre des Prêcheurs (1215–1237)* (Fribourg, 1977).

RIGON, ANTONIO, 'Chiesa e vita religiosa a Padova nel duecento', in *S. Antonio 1231–1981: Il suo tempo, il suo colto, e la sua città* (Padua, 1981), 284–99.

—— 'Francescanesimo e società a Padova nel duecento', in *Minoritismo e centri veneti nel duecento nell'ottavo centenario della nascita di Francesco d'Assisi (1182–1982)* (Civis studi e testi, 7; Trento, 1983).

—— 'Vescovi e ordini religiosi a Padova nel primo duecento', in *Storia e cultura a Padova nell'età di sant'Antonio* (Fonti e ricerche di storia ecclesiastica padovana, 16; Padua, 1985), 131–51.

ROSA, ITALO, *Il beato Giordano Forzaté, abbate e priore di San Benedetto in Padova, 1158–1248* (Bresseo, 1932).

RUSCONI, R., 'Predicatori e predicazione', in *Annali della storia d'Italia*, 4 (Turin, 1981), 949–1035.

SALVEMINI, GAETANO, 'La lotte fra stato e chiesa nei comuni italiani durante il secolo XIII', in *Studi storici* (Florence, 1901), 39–90.

SANDRI, GINO, 'Paquara e Vigomondrone', *Atti dell'Academia di Agricoltura, Scienze e Lettere di Verona*, 5th ser., 13 (1934), 101–15.

SCHEEBEN, HERIBERT CHRISTIAN, 'Prediger und Generalprediger im Dominikanerorden des 13. Jahrhunderts', *AFP* 31 (1961), 112–41.

SCHIO, GIOVANNI DA, and SCHIO, ALVISE DA, *Fra Giovanni da Vicenza a Paquara: 28 agosto 1233–1933* (Schio, 1933).

SEVESI, PAOLO, 'Beato Leone dei Valvassori da Perego dell'Ordine dei Frati Minori, arcivescovo di Milano (1190?–1257)', *Studi francescani*, 2nd ser., 13 (1927), 70–93; 14 (1928), 44–55.

SIGAL, PIERRE-ANDRÉ, *L'Homme et le miracle dans la France médiévale (XIᵉ–XIIᵉ siècle)* (Paris, 1985).

SILINGARDI, GIANCARLO, and BARBIERI, ALBERTO, *Storia di Reggio Emilia* (Modena, 1970).

SILLI, A., 'B. Giordano da Rivalto', *Biblioteca Sanctorum* (1964).

SIMEONI, L., 'Le origini del comune di Verona', *Nuovo archivio Veneto*, 25 (1913), 49–143.

—— *Il comune veronese sino ad Ezzelino ed il suo primo statuto* (Miscellanea di storia Veneta, 3rd ser., 15; Venice, 1922).

SIMIONI, ATTILIO, *Storia di Padua dalle origini alla fine del secolo XVIII* (Padua, 1968).

SISMONDI, J. C. L., *History of the Italian Republics in the Middle Ages*, ed. William Boulting (London, 1906).

STARN, RANDOLPH, *Contrary Commonwealth: The Theme of Exile in Medieval and Renaissance Italy* (Berkeley, Calif., 1982).

STEFANO, ANTONIO DE, 'Le origini dei Frati Gaudenti', in *Riformatori ed eretici del medioevo* (Palermo, 1938), 211–69.

SUTTER, CARL, *Johann von Vicenza und die italienische Friedensbewegung im Jahre 1233* (Freiburg, 1891); trans. Maria, Gelda, and Olga da Schio, *Giovanni da Vicenza e l'Alleluja del 1233* (Vicenza, 1900).

SYMONDS, J. A., 'Religious Revivals in Medieval Italy', *The Cornhill Magazine*, 31 (1875), 54–64.

THOMPSON, AUGUSTINE, 'Le tentazioni del predicatore nella vita di Ambrogio

Sansedoni', *Bollettino di S. Domenico*, 67 (1986), 145–50.

THOMPSON, WILLIELL R., *The Friars in the Cathedral: The First Franciscan Bishops, 1226–1261* (Toronto, 1975).

THOUZELLIER, CHRISTINE, 'La Légation en Lombardie du cardinal Hugolin', *Revue d'histoire ecclésiastique*, 45 (1950), 508–42.

TIRABOSCHI, GIROLAMO, *Storia della letteratura italiana* (8 vols.; Venice, 1823–5).

TONDELLI, LUIGI, 'L'anno del'"Alleluia"', in *Il Libro delle figure dell'Abate Gioachino da Fiore* (Turin, 1953), i. 192–6.

TOURON, ANTOINE, *Histoire des hommes illustres de l'Ordre de saint Dominique*, (6 vols.; Paris, 1743–9).

TURNER, VICTOR, *Dramas, Fields, and Metaphors: Symbolic Action in Human Society* (Ithaca, NY, 1974).

—— *The Ritual Process: Structure and Anti-Structure*, (Ithaca, NY, 1982).

—— and TURNER, EDITH, *Image and Pilgrimage in Christian Culture: Anthropological Perspectives* (New York, 1978).

VALSECCHI, FRANCO, *Comune e corporazione nel medio evo italiano* (Milan, 1949).

VAUCHEZ, ANDRÉ, 'Une campagne de pacification en Lombardie autour de 1233: L'Action politique des ordres mendiants d'après la réforme des statuts communaux et les accords de paix', *EFRMAH* 78 (1966), 519–49.

—— *La Sainteté en occident aux derniers siècles du Moyen Âge d'après les procès de canonisation et les documents hagiographiques* (Rome, 1981).

VERCI, GIOVANNI BATTISTA, *Storia della Marca trevigiana e veronese* (20 vols.; Venice, 1786–91).

VERGOTTINI, GIOVANNI DE, 'Il "popolo" di Vicenza nella Cronaca ezzeliniana di Gerardo Maurisio', *Studi senesi*, 48 (1939), 354–74.

—— *Arti e popolo nella prima metà del secolo XIII* (Milan, 1943).

—— 'Note sulla formazione degli statuti del popolo', *Revista di storia del diritto italiano*, 16 (1943), 61–70.

—— *Studi sulla legislazione imperiale di Federico II in Italia: Le leggi di 1220* (Milan, 1952).

Vescovi e diocesi in Italia nel medioevo secoli IX–XIII (Atti del II Convegno di Storia della Chiesa in Italia; Italia Sacra 5; Padua, 1964).

VICAIRE, MARIE-HUMBERT, *Dominic et ses prêcheurs*, 2nd edn. (Paris, 1977).

VODOLA, ELISABETH, *Excommunication in the Middle Ages* (Berkeley, Calif., 1986).

VOLPE, GIOCCHINO, 'Liber Maiolichinus', *Archivo storico italiano*, 5th ser., 37 (1906), 93–114.

—— *Medio evo italiano* (Florence, 1961).

WALEY, DAVID P., *The Italian City-Republics* (New York, 1969).

WALTER, I. 'Benedetto', *Dizionario biografico degli italiani* (1966).

WALZ, ANGELUS MARIA, *Compendium Historiae Ordinis Praedicatorum* (Rome, 1930).

WARD, BENEDICTA, *Miracles and the Medieval Mind: Theory, Record and Event, 1000–1215* (Philadelphia, 1982).

WEINSTEIN, DONALD, and BELL, RUDOLF M., *Saints and Society: The Two Worlds of Western Christendom, 1000–1700* (Chicago, 1982).

WEST, DELNO C., jr., 'The Education of Salimbene of Parma: The Joachite Influence', in Ann Williams (ed.), *Prophecy and Millenarianism: Essays in Honour of Marjorie Reeves* (Harlow, 1980), 193–215.

—— and ZIMDARS-SWARTZ, SANDRA, *Joachim of Fiore: A Study in Spiritual Perception and History* (Bloomington, Ind., 1983).

WINKELMAN, EDUARD, *Kaiser Friedrich II* (2 vols.; Leipzig, 1889–97).

INDEX

Adriani, Giavombatista 191
Albert, bp. of Ceneda 65, 76, 79
Alberti, Leandro, OP 209
Albigensians 123
Alexander VI, pope 12
alliances, political 8–9, 41, 51, 63–4, 77
  see also individual cities
Almerico of Bologna 79
Ambrose Sansedoni, OP:
  in art 126
  on John of Vicenza 112
  as peace-maker 150–6
  as preacher 11, 17, 102, 135
  sermons 22, 102–3
  statute reformer 199
Ancient Way of Preaching', the 15–18,
  20–3, 25
ancients 8, 48, 139
  see also communes; popolo
Andito, William De 36
Anthony of Padua, OFM, St:
  his audience 85, 92, 98–9, 105
  canonization 42, 72
  cult and relics 22, 130
  miracles 90, 108, 120–1, 125, 126–7,
  213
  and peace-making 152
  and usurers 185
anti-Semitism 185 n.
Anzola 49, 61
appeal of peace arbitration 165–6
Apulia 32, 209
arbiters 157–78
arbitration decrees 168–73, 172–7
  see also mediation
agreement to arbitration, see compromissum
Arezzo 150
Ariosti, Gerard 45
Armanno of Porta Nuova 56, 146, 162,
  181
Armi, Società delle:
  in Bologna government 47, 215
  and John of Vicenza 54–5, 99, 107
  origins and organization 6–8, 201
  and peace-keeping 140, 201
  see also popolo
Arti, Società delle 6–8, 47, 215
  see also popolo

Assisi 149
audience of preaching 83–5, 91–101
ban:
  against moral offenders 188
  against peace-breakers 175–6, 204
  relaxation by revivalists 159–60, 164,
  202–4
  see also sanctions for peace-breaking
Barnabas of Reggio, OFM 15–16, 90–1,
  106
Bartholomew of Breganza, OP 16, 21–2,
  38, 128, 161, 217
Bartholomew of Trent, OP 96, 104, 124,
  126, 131
Bartholomew of Verona, OP 73
Bassano 66
Belluno 7, 74
Benedict (Cornetto), br:
  precursor of Alleluia 29–32
  style of preaching 29–30, 43–4, 93,
  100, 154
Benedictines 179
Bergamo 25, 34, 51, 73, 123, 192, 209
Bernardino of Siena, OFM 12, 25, 218
Berthold of Regensburg, OFM 13, 91, 92,
  98, 105, 218
Bertold, abp. of Aquileia 73
betrothal, see marriage
Bigari, Vittorio 57 n., 132
Blandrato, Ardicino 191
blasphemy statutes 185, 189
Boccabadati, Gerard, see Gerard of Modena
Bologna:
  base for preaching by John of Vicenza
  67–9, 71
  conflict between commune and bishop
  45–50, 162–7, 170–1
  Francis of Assisi at 85
  internal politics 3, 5–9, 140–1
  John of Vicenza at 45–62, 93–4, 99,
  107, 131, 154, 210
  statute reform 172, 175, 180–1, 185,
  192, 200
Bologna, University of 46, 49, 84, 97
Bolognitto of Strada Maggiore 148
Boncambio, James, OP, bp. of Bologna
  73, 97–8, 214

Bonatti, Guido 96, 102, 107, 112
Boncompagno of Signi 84, 102
Bono, Vitale de 71
Borselli, Gerolamo de', OP 19, 159, 187, 219
Brescia 64, 65, 67, 70, 73–4, 79
Buonviso of Bologna, OP 87

Camposampiero family 64, 65, 67
  Tisone 76
Captain of the People 8
  see also *popolo*
Castel d'Argile 61
Castel del Vescovo 61
Castel San Pietro 55
Castelfranco 51, 54
Castello del Vescovo 49
Catalonia 157
Cathars 193
Ceneda 65, 76
Cento 50
Cesena 101, 108
chastisement miracles 114–17, 131–5
Church, statutes on the 182–4, 189–200
Cistercians 179, 217
Claro of Florence, OFM 84
commissioning of mediators 160–7
communes 1–9
compromissum 161–5
Como 38, 133, 216
Conegliano:
  John of Vicenza's arbitration for 76,
    78–9, 162, 166, 206
  in Veneto politics 64–7
confession 131
conflict resolution 138–43, 145–50,
    173–4
  see also mediation
confraternities 217
Conrad of Metz 47
consorterie 6
Constance, Peace of 2, 47
Constitutio in Basilica Sancti Petri 191
  see also heresy; heretics
contraceptives 188
Cornetto, br, see Benedict (Cornetto), br
Cremona 5, 51, 60, 65
Cronaca Rampona 99, 159, 219

Da Camino family 64–6, 67, 70, 74,
    76–7, 79
Da Romano family:

Adelaide 75
Alberico 63–6, 75, 76
Ezzelino 63–6, 70, 74, 75, 98, 132,
    205, 213
  in Veneto politics 63–6, 67, 69, 72
  John of Vicenza favors 77–9, 206–8
  at Paquara 74–6
della Fratta, Henry, bp. of Bologna:
  as bishop 45–7, 51
  conflict with commune 45–9, 50, 140,
    147, 162
  at St Dominic's translation 59
d'Este family:
  Azzo 63, 65, 66, 74
  Rinaldo 75
Detesalve of Florence, OFM 129
the Devil 119, 124, 127
Diet of Roncaglia 2
Dominic, St:
  canonization 58–9, 96, 111
  and preaching 11, 87, 89, 123
Dominican General Chapter (1220) 41
Dominican General Chapter (1233) 52, 58,
    60, 146, 210
Dugliolo 48–9, 61

Elias of Cortona, OFM 16, 124, 146
Este, see d'Este
excommunication 48–50, 65, 69, 72, 75,
    104, 134, 148–9, 164, 177
exile 141, 145, 159, 160
exorcism 104, 121, 124

Faenza 42, 55, 65, 124, 154
family law 198–9
fear as element of preaching 104–6, 131
Feltre 67, 74
Ferrara 29, 61, 69, 70, 74
Ferri family 191
feuds, settlement of 145–6, 150–1,
    157–8, 173
Fiesso 61
Fiorenzuola 37
Fioretti 137–8
flagellants 10 n., 21, 33
Florence 12, 58, 69, 84, 88, 119, 127,
    149
folklore as a source 16–18
Fontanella 37
force, use by revivalists 106–8
Forlì 55
Forzaté, Jordan, see Jordan Forzaté

Foulques of Neuilly 17, 83–4, 99, 104, 107, 152
Francis of Assisi, St 85, 90, 98, 127, 137, 149–50
Frederick I Barbarossa, e. 2, 47
Frederick II, e.:
    and Alleluia preachers 32, 36, 100, 104, 111, 207
    and heresy 190, 191, 192, 195
    and north Italian politics 2, 8, 47, 48, 64
Frederick of Lavellongo 48–9, 148
freedom of the Church, statutes on 182–4, 196–9
Fumagalli, Vito 153, 216

gambling 185, 186
Gandulf of Gisso 56, 146, 181
Genoa 3
Gente, Ghiberto de 43, 142
Gerard of Modena, OFM:
    and Alleluia 33–6, 39, 42
    family 44
    miracles 35, 54, 89, 129
    peace-making 16, 34, 35, 124, 142, 145, 159–60, 162, 171, 176
    statute reform 56, 145–6, 159, 163, 164, 176, 180–204, 214
'Grace of Preaching', the 88–91, 127
Gratian of Padua, OFM 119, 212
Gratian the Canonist 157
Grazia, bp. of Parma 49
Gregory IX, pope:
    and John of Vicenza 42, 58, 59, 68, 72, 76, 78, 112, 126, 207, 210, 211
    as legate in Lombardy 192
    and north Italian politics 34, 49, 191
Guala, OP, bp. of Brescia:
    Alleluia preacher 38, 44, 73, 209
    as peace-maker 34, 35, 51
    vision of St Dominic 123
guardianship 171
Gubbio 137
guidardonum 186
Guido, bp. of Vicenza 64
Guido, OP, prior of Padua 41
Guido of Rho 65
Guido of Sasso 5
Guidotto, bp. of Mantua 49

healing, miraculous 108, 115, 116, 121, 219

hell 104, 105
Henry of Cominiciano, OFM:
    legislation against heresy 181, 190–2, 194, 196, 214
    not a preacher of Alleluia 181–2, 191–2
    statute reform at Vercelli 180–1, 185
Henry V, e. 2
heresy, repression of:
    confraternities 217
    imperial legislation 191–3
    municipal statutes 181–2, 189–96
    not dominant element of Alleluia 196
heretics:
    attempts to convert 133, 218
    murder Peter of Verona 208
    persecution of 38, 71, 104, 190, 195, 206
Holy Spirit and preaching 126, 131
homicide 143–5, 148, 158, 160, 169–70, 172, 204
Hugh of Saint-Cher, OP 122
Hugo, bp. of Vercelli 191

Imola 55
indulgences 69, 208, 211
Innocent III, pope 45
Innocent IV, pope 34, 208
Isnard of Chiampo, OP 115–16, 128, 133
Ivo of Chartres 136

Jacopino of Reggio, OP:
    education 34, 44
    miracles 35, 111, 129
    preacher of Alleluia 16, 33, 43, 99
    death 129, 209
James of Bussolaro 11
James of Faenza, OP 124
Jews 185 n.
Joachim of Fiore 11, 23 n., 122, 136 n.
John di Biagio 108
John of Vercelli, OP 125, 209
John of Vicenza, OP:
    after 1233 79, 203–9
    arbiter at Bologna 56–61, 140, 145–8, 166, 170–1, 209–10
    in art 57 n., 59 n., 124 n., 126, 131
    his audience 55, 67–9, 92–6, 128–9
    Bologna preaching campaign 33, 42, 52–62
    and canonization of St Dominic 59, 111–12, 208
    devotion to 55, 128–9

John of Vicenza, OP: (*cont.*)
  historiographical treatment 17–20, 41
  miracles 35, 54, 84, 94–7, 112–14,
    120, 132, 208–9
  at Paquara 72–6, 152, 154, 174–5,
    177
  peace-making 56–60, 66–71, 72–6,
    145–8, 161–7, 168, 174–5, 177,
    202–3
  as preacher 12, 52–6, 60–1, 70–1,
    74–5, 87–8, 92–7, 102, 106–7,
    122–3, 125, 130, 134, 152
  and politics of Bologna 52–7, 99,
    106, 132, 154, 159
  and politics of Veneto 41, 71, 75–9,
    200–1, 213, 215–16
  statute reforms 57, 77, 180–4, 185–7,
    200
  Veneto preaching campaign 66–80,
    203–7
  youth to 1233 39–42, 44
Jordan Forzaté, OSB:
  opposition to John of Vicenza 72, 76,
    79, 101–2, 207, 210
  political activities 42, 63, 74, 76
Jordan of Pisa, OP 86
Jordan of Saxony, OP 42, 59, 60, 102,
  127, 146, 149, 210

Kee, Howard Clark 117–18
kiss of peace 158–9, 170–4

Landolfo, Pasquale 53, 106, 185
Languedoc 217
Lateran Council III 157
Lateran Council IV 45, 141
law, canon 148, 157, 186
law, feudal 147–8
law, Roman 2, 46, 148, 165, 180
Lea, Henry C. 181, 189, 196
legal aid 197–9
Legend of Perugia 149–50
Legislation on preaching 86–8, 103, 127
Leo de' Valvassori, OFM, abp. of Milan:
  career 16, 36, 44, 209
  peace-making 37
  statute reforms 180–1, 193
  visions 89, 105, 128
license to preach 127–8, 160
liminality 93, 155 n.
Lombard faction 64–5, 70, 79, 207
Lombard League, First 2
Lombard League, Second 2, 47

Lorenzo di Pietro (Il Vecchietta) 126
Luke the Lector, OFM 122

Mainerio, Lantelmo 37
Malaguzzi, Sigismund 43
Maletta, *see* Gerard of Modena
Manelino 39
Mantua 64, 65, 67, 70, 73, 74, 129, 192
marriage, role in peace-making 75, 174,
  187 n., 198–9
Massumatico 49, 61
Matthew, syndic of Conegliano 76
Matthew Paris, OSB 98
Maurisio, Gerardo 7, 20, 77, 79, 95–6,
  205, 215
mediation 157–67
Milan 38, 47, 65, 85, 104, 180, 192
Milancio of Bologna 72
Militia of Jesus Christ 110, 217
minting 3
miracles and miracle stories:
  collections 96, 114
  interpretation 21–2, 111–17
  and revival preaching 16–17, 21, 84,
    90, 94–6, 107–9, 110–11, 118–26,
    158, 213
  during sermons 96, 120–3
  types and significance 114–35
Modena 24, 43, 51–4, 60, 93, 203, 216
Moneta of Cremona, OP 97
Monselice 62, 66
Montecavalloro 49, 61
Montecchi faction 63, 65, 70, 77
Monza 25, 38, 180–1, 192–3
moral reform legislation 185–9, 199
murder, *see* homicide

Namur 153
Nicholas, bp. of Parma 44
Nicholas of Dacia 134
Nicholas of Doara 43
Norbert of Xanten, St 122, 153–5
notarial manuals 161–5

oaths 153, 164
  see also perjury
Odofredus of Bologna 180
Olremo 65
Opeano 65
the ordeal 142
orphans, *see* widows and orphans
Ostiglia 65, 79
Otto, papal legate 65

Otto IV, e. 45
Ozzano 61

Padua:
    internal politics 70, 73–6, 85
    statutes 25, 143–4, 180–2, 192
    in Veneto politics 62–6, 77–8, 206
Padua, University of 50
Pagliarini, Giambattista 205
Palmerio of Santa Trinità 49
Paquara, Peace of 68, 71–6, 92–3, 162,
    174–7, 207, 216
Parma:
    peace-making at 145, 146, 164, 200–4
    in regional politics 51, 54
    revivals at 18, 34, 60
    statute reform 24–5, 142, 168,
        180–204, 216
Passagerii, Rolandinus, *see* Rolandinus
    Passagerii
Patarins 195
Pavia 11, 128, 133
the Peace and Truce of God 157
peace pacts 137–8, 142, 145–50,
    159–78
peace-breaking 145, 151, 160, 171–8,
    201–4
peace-making 136–56
    during Alleluia 23, 142, 216
    legal forms 157–78
    and penance 137, 151–4
    and preaching 150–6
    statutes on 139–43, 170–4, 175–8
penance and revivalism 56, 74, 107, 153
perjury 148, 160, 204
    see also oaths
Peter of Verona, OP, St:
    and Alleluia 17, 212
    audience 22, 84–5, 100, 129
    and heretics 133, 208, 217
    miracles 108, 115, 119–22
    murder of 208
    as preacher 95, 127, 131, 211
    statute reform 180
Philip, bp. of Ravenna 208
Piacenza 6, 18, 36–8, 42, 87, 192, 213
Pianio 65
Pieve di Cento 50
Pilio, Uguccione da 64, 77–9
Pirovali, Guifredo de' 36
Pisa 3
pledges in peace-making 175
podestà 5–6

Poggetto 61
popolo, the 3, 7–9, 36, 40, 47, 99
Prato della Valle 68, 85
prisoners, release of 58, 67, 159, 203
processions 32, 54–8, 93–4, 100, 107,
    154–5
prophecy 122–6, 133
prostitutes 104, 181, 187–8
publicity, diffusion of 94–8
pulpits, movable 92, 94
Pusterla, William 40–1

Quatuorviginti 63, 65, 70, 77

Rainerius Perusinus 161
Rambertini, Uguccio dei 40
Reggio 33, 42–3, 49, 51, 148, 216
Reginald of Orleans, OP 97
relics 107, 129–30
Remigio de' Gerolami, OP 203
Reni, Guido 124 n.
Richard of San Germano 31
Richardus Anglicus 161
rituals of peace-making 151–6, 158,
    172–4
Rivoli, Henry de' 77
Robert of Arbrissel 136
Robert of Modena 79
Roccaforte 37
Roland of Cremona, OP 37, 97, 102, 213
Rolandino of Padua 20, 67, 152, 208
Rolandinus Passagerii 24, 161, 163–7,
    173, 174–5
Rome 5, 68, 88
Rosano 37
Rossi, Bernard dei 34
Rovigo 61, 65
Ruggieri, Hippolytus 43

Salimbene, OFM:
    on the 'Ancient Way' and the Alleluia
        15–16, 23, 31–2, 211–12
    on miracle-working 21, 94, 110
    on preachers 16, 31–2, 41, 53, 88,
        90–1, 105, 129, 191, 209
    peace-maker 149
    as source 15–16
Salvolini, Alessandro 124 n.
San Bonifacio family 70, 78, 141
    Richard 64–5, 66–7, 70, 74–5, 79
San Cesario 51
San Germano 32
San Germano, Peace of 48
San Giovanni in Persiceto 45, 49, 61, 148

San Giovanni Lupatoto 73
San Jacopo della Tomba 73
sanctions for peace-breaking 171–2,
    175–8
Sansedoni, Ambrose, *see* Ambrose Sansedoni
Sant'Ilaria 43
Savonarola, Jerome, OP 12, 25, 212, 214,
    218
scapegoating 102–3
Siena 12, 58, 69, 151–4, 214
Sigal, Pierre-André 116–18
sorcerers 50, 187–8
Speranza, Giovanni 131
Spoleto 29, 31
statutes, municipal:
    reformed by revivalists 57, 179–204,
    214
    as sources 18, 25, 179–84
Stephen of Spain, OP 58, 111, 190, 196
sumptuary laws 53, 187
Sutter, Carl 10, 18–19, 189, 196, 205,
    216

Tancred of Bologna 161
Tealdisco, Guizardo 70
Theobald of Albinga, OP 132, 150,
    158–9, 167–8, 169, 172–3, 212
Theodoric, abp. of Ravenna 59
Thomas of Cantimpré, OP 96, 113
tithes 45, 48, 50, 145, 147
Tonisto, Nicholas 79
torture 193–5
training of preachers 86–7, 104
Traverse di Porta S. Procolo 140
Treviso 63–6, 70, 76, 79, 144, 179, 192,
    194, 196
the Tyrol 65

Ugolino, cardinal, *see* Gregory IX, pope
usurers and usury 50, 53–4, 104, 106,
    131, 133, 152, 181, 185–6

Vauchez, André 180, 181, 182–4, 189,
    192, 196, 205, 216
Valvassori, Leo de', *see* Leo de' Valvassori
Vercelli statutes 25, 38, 145, 180–1,
    185–6, 190–2, 194, 204
Verona:
    internal politics 47
    John of Vicenza and 70–2, 77–9, 131,
    190
    statutes 25, 140–1, 144, 173, 180,
    194–6, 201–2
    in Veneto politics 41, 63–6
Vicenza:
    internal politics 7, 47
    John of Vicenza and 39–41, 70–2,
    77–9, 108, 206
    statutes 25, 139, 144, 180, 201–2
    in Veneto politics 64–6
Visconti, Federico, OFM, abp. of Milan
    89 n., 98
Visconti, Galeazzo 12
visions 89, 94, 105, 117, 123–4
the *Vitae Fratrum* 150, 219
Viterbo 72
Vivaresi faction 40–1

widows and orphans 56, 101, 146, 197–9,
    214
William, bp. of Modena 61, 68–9, 73,
    126
William of Cordella, OFM 121
William of Cremona 65
the 'Wolf of Gubbio' 137–8
women:
    and revivalists 53, 91, 100–1, 197–9
    statutes on 187–8, 197–9, 204

Zachame family 77
Zeno, Rainier 49